Darwin's Proof

Darwin's Proof

The Triumph of Religion over Science

Cornelius G. Hunter

Brazos Press
A Division of Baker Book House Co
Grand Rapids, Michigan 49516

© 2003 by Cornelius G. Hunter

Published by Brazos Press
a division of Baker Book House Company
P.O. Box 6287, Grand Rapids, MI 49516-6287
www.brazospress.com

Printed in the United States of America

All rights reserved. No part of this publication may be reproduced, stored in a retrieval system, or transmitted in any form or by any means—for example, electronic, photocopy, recording—without the prior written permission of the publisher. The only exception is brief quotations in printed reviews.

Scripture is taken from the HOLY BIBLE, NEW INTERNATIONAL VERSION®. NIV®. Copyright © 1973, 1978, 1984 by International Bible Society. Used by permission of Zondervan. All rights reserved.

Library of Congress Cataloging-in-Publication Data is on file at the Library of Congress, Washington, D.C.

Contents

Preface   7

1. Darwin's Deceptive Idea   9
2. Swallowing a Camel: The Fundamental Argument against Evolution, Part 1   14
3. Swallowing a Camel: The Fundamental Argument against Evolution, Part 2   26
4. Straining at the Gnat: The Scientific Evidence for Evolution, Part 1   36
5. Straining at the Gnat: The Scientific Evidence for Evolution, Part 2   50
6. Blind Guides: The Philosophical Argument against Evolution   65
7. Another Gospel: The Theological Argument against Evolution   82
8. A Reason for Hope: The Only Explanation for Life   97
9. What Has Been Made: One Long Parable   109
10. Come Let Us Reason: The Intelligent Design Theory   116
11. Wisdom Rejoices: The Nature of Design   126

Appendix: Faulty Arguments for and against Evolution   135
Notes   155
Index   165

Preface

If it is true that to err is human then science is very human. From alchemy to radium tablets, science has a long history of blunders. But science learns from its mistakes and it is better known for its great triumphs. The mistakes are just part of the process. Scientists must watch for them and weed them out. Hence an important part of science is its constant introspection. It must always evaluate and reevaluate what it thinks is true. Scientists are taught that nothing is sacred—even the most popular theories may well be a mistake.

Introspection can be a difficult thing. It is far easier to see the blunders of the past than those of the present. Mistakes in the history of science can be analyzed dispassionately, but there is more at stake in contemporary thought. This book is about Charles Darwin's theory of evolution. It is, I believe, a blunder but a complicated one. Evolution is not good science, but this is not the whole story. It is not enough merely to point out how the scientific evidence is at odds with Darwin's theory. For evolution is an explanation of our origins, and inevitably the subject of origins involves religious sentiment. Ultimately, to understand evolution, we must understand the religious ideas that feed it.

This book examines both the history and impact of those religious ideas. The story is complicated because it involves a subtle and deceptive mixture of fact and assumption. The facts are the observations of biology and the natural world; the assumptions are beliefs about how the world should work and about the actions of God. It can be exceedingly difficult to tease these facts and assumptions apart. The powerful arguments for evolution may sound scientific but there are imbedded crucial metaphysical premises.

There is a great consistency to evolutionary thought. The same rationale is used over and over. The reasoning and mode of argumentation may be subtle, but once it is understood it becomes obvious—a common thread running through the arguments for evolution. And once it is understood our view of evolution changes dramatically. It is not a scientific fact or even compelling theory. It is a religious philosophy that has found a home in science and dictates the underlying assumptions. It is still true, as it was said long ago, that theology is the queen of the sciences.

Finally, this book presents a better way to understand biology and life's origins. It will always be rejected by evolutionists for it is not based on evolution's religious philosophy. Again, despite the obvious evidence at hand, evolutionists are guided by the religion at the core of their theory.

In writing this book I am indebted to a great many people. In addition to my parents and wife Jeanine, I would like to thanks Ashby Camp, Joshua Coe, Stephen Jones, Kirk Durston, J. Brian Pitts, Kenneth Harris, Dean Orr, Michael Shea, and Phillip Johnson. I am responsible for any errors in the book.

1

Darwin's Deceptive Idea

Charles Darwin proposed his theory of evolution in 1859. Darwin said that all the species arose naturally. He claimed that populations in the wild are far more fluid than they appear to be. Whereas we observe distinct species, Darwin believed that over long periods of time those species could gradually evolve to become something completely different. A fish, for example, could become a frog. Thus the species were not created as they are, but rather arose gradually via natural processes.

No one doubts that species adjust to changing conditions. It is a fascinating and critical part of biology. The beaks of birds, for example, have been observed to change slightly during drought years. But the beaks are still beaks and the birds are still birds. Darwin's theory of evolution is not that such modest levels of change are possible, but that they somehow add up to far greater levels of change.

So while everyone agrees that evolution occurs to a very limited degree, what remains unobserved and downright questionable is Darwin's claim that it created all the species. Nonetheless, practically since

Charles Darwin first proposed his theory, evolutionists have insisted that the controversial idea is a fact. The details may not all be known, they allow, but the overarching idea is not in question.

In the decades following the publishing of Darwin's works, Berkeley professor Joseph Le Conte argued that those who believe in evolution should not be referred to as evolutionists any more than those who believe in gravity should be referred to as gravitationalists. Evolution, according to Le Conte, was not to be regarded as a mere theory, but as an unquestionable fact, or scientific law, on par with gravity.

Today, evolutionist Ernst Mayr echoes Le Conte's message that evolution should no longer be called a theory. Professor emeritus of zoology at Harvard University and one of the twentieth century's foremost evolutionists, it would be difficult to find a more authoritative voice in Darwinism than Mayr. The fact of evolution is "so overwhelmingly established," says Mayr, "that it has become irrational to call it a theory."[1]

The idea that evolution is a fact beyond rational dispute is broadly popular among Darwinists. In the hundred years or so separating the comments of Le Conte and Mayr, many evolutionists have made a similar claim. It is now seen as accepted wisdom in biology textbooks and popular literature.

The Fact of Evolution

This is a curious argument since the scientific evidence does not establish the fact of evolution. The evidence cited in favor of evolution does not hold up well under scrutiny, and there is plenty of evidence against evolution. In my previous book, *Darwin's God: Evolution and the Problem of Evil*, I reviewed all the major evidence that evolutionists say proves their theory to be a fact. In every case the evidence is ambiguous or even argues against evolution. How then can evolution be a fact if even the positive evidence does not support it very well? The answer is that evolution is considered to be a fact because Darwinists believe they have disproven the alternative: divine creation. I showed how the evidences are repeatedly interpreted as powerful arguments for evolution by virtue of their implications for divine creation.

With divine creation, evolutionists point out, we must believe, for example, that God created species that later became extinct, that God created species with many similarities, and that God created a world with parasites and other dangers. I showed that these and many other arguments are consistently used by evolutionists when arguing for their theory. I also showed that this sentiment was popular in the years leading up to Darwin. I gave examples of traditions within the church that antici-pated and even motivated Darwinism.

As early as the seventeenth century, almost two hundred years before Darwin published his theory of evolution, philosophers and theologians were calling for an intermediate process between God and creation. By the time Darwin came around this call had only increased in volume. What was needed was a credible explanation of how the process worked. This was Darwin's contribution, for he filled in the details. With Darwin a rather ill-defined tradition, or set of traditions, became formalized.

But this was only part of the story. It was not merely a happy coinci-dence that Darwin's theory enjoyed support from certain areas outside of science. It was not serendipity that science was now proclaiming what many thinkers had envisioned in the preceeding centuries. The other part of the story was that Darwin's development of the theory, as well as its continuing success, hinged on these sentiments about God and creation.

The Religion in Evolution

My conclusion in *Darwin's God* was not that evolution lacks evidence or that it is false. It has plenty of strong arguments but they rely on a cer-tain view of God and creation. There is no way to reformulate the argu-ments without this key premise except by stripping them of their force, but then evolution would no longer be a fact; it would not even be very likely.

Is it not true, some will ask, that all of science incorporates religious assumptions? After all, physicists and chemists assume that there is a uni-formity behind nature yet they cannot prove this to be true. This assump-tion must ultimately derive from a religious belief. This is true but not relevant to Darwinism. Modern science does include certain unprovable assumptions such as uniformity and parsimony, but Darwinism relies on more assumptions than just these. In particular, Darwinism's view of God and creation would be difficult to reconcile with Christianity.

Darwinism depends on religion, but only to overrun the opposing the-ory. Once this work is done evolutionists are free to pursue an entirely mechanistic explanation of life. Evolution, by default, becomes the explanatory filter for all we observe in nature, no matter how awkward the fit. About two thousand years ago the great mathematician and astronomer Ptolemy developed his complicated explanation for his Earth-centered system. He added epicycle upon epicycle to explain the motions of the Sun, Moon, and the planets. He could even predict eclipses.

So it is with Darwinism. Evolutionists add layer upon layer of circuitous explanation to fit nature into their theory. It sounds scientific because the explanation is purely mechanistic. There are no religious claims, for example, in the technical research journals, but journal articles do not attempt to prove the theory. They attempt to explain how evolution must have occurred, assuming that it did occur.

All of this means that Darwinism is subtle. It is not merely a case of science versus religion, or a case of atheism making its way into science. Nor is it a case of controversial religious ideas undermining science. The religious claims of Darwinists are, for many people, quite reasonable. They may not square with Scripture, but they sound good. Indeed, Darwinists usually state their claims as though they were simply a matter of fact. The religion that fed into Darwinism is taken for granted.

In the past half-century or so it has become fashionable to see science as a social construct driven by politics, funding constraints, personalities, and so forth. While such influences are obviously a reality in science, I do not believe they are usually dominant. Science can and has transcended outside influences to produce new and unique findings, but Darwinism is not one of them. Darwinism is an example of how powerful and yet subtle outside influences can steer our scientific thinking.

It was a sufficient task for me in writing *Darwin's God* to reveal these subtleties and expose the true nature of Darwinism. My aim was not to provide an alternative or to argue that evolution was wrong, therefore I did not detail the many evidences against the theory nor provide a replacement theory. I also did not discuss my personal religious belief, as it was not relevant to the task of critiquing Darwinism.

An Alternative Explanation

In this book I will argue that Darwinism is wrong, and I will present an alternative view. I will argue that Darwinism is wrong on four levels. First, in chapters 2 and 3 I will detail a sampling of the evidences against evolution. These are some of the many examples of complexity in biology that defy naturalistic explanations. It is always possible to contrive naturalistic explanations for what we observe in biology, but we must ask how likely such explanations are. On what basis should we believe that the most complex of machines that we know of arose on their own? As we shall see in these chapters, the Darwinist account is long on speculation and short on compelling explanations.

Chapters 4 and 5 argue against Darwinism on the second level. They show that even the positive evidences do not support evolution. Chapter 6 shows how Darwinism fails on yet another level: it is self-contradictory. Finally, Chapter 7 argues against Darwinism on the theological level.

Darwinism fails scientifically, philosophically, and theologically. It is not a good creation story. Chapters 8 and 9 argue that the biblical account is the only one that makes sense. It may not provide many scientific details, but it accurately describes the situation with which we are faced. As a Christian I know what the Scriptures say, and as a scientist I understand what we have discovered about nature. These two sources of revelation tell the same story. It is a myth that Christians must, in one way or another, reconcile their religious beliefs with what science has discovered.

How then should Christians conduct scientific research, especially in the life sciences? In Chapters 10 and 11 I argue for the intelligent design (ID) framework. Some argue that ID is not a very good Christian apologetic. Others argue that ID doesn't provide enough details—it doesn't tell us how the species were created or the age of the earth. Both criticisms are correct. ID is not an argument for the Bible, and it is not a theory of everything. It does, however, make predictions and provide a framework for scientific research.

One common criticism is that ID is a religious theory by virtue of its appeal to a designer. As we shall see, it is evolution rather than ID that makes strong religious assumptions. ID is unacceptable to Darwinists because ID does not incorporate evolution's religious assumptions, but ID has no problem with the existence of particular evolutionary processes to the extent that they are supported by the empirical evidence. This is one reason why ID allows for more diversity in its explanation than evolution. Darwin's exclusive focus on natural laws fails to account for our complex and interconnected biological world.

Another failed criticism is that ID is a stop-gap explanation that depends on our lack of knowledge. According to these critics, ID is merely a label used to explain things we have not yet figured out about biology. Nothing could be farther from the truth. The design inference is based on empirical evidence and rational analysis, not ignorance. Again, we shall see that it is evolution, rather than ID, that is a stop-gap explanation. Evolution loses its explanatory power as our knowledge increases. ID makes scientific predictions and provides a framework upon which to formulate subhypotheses and pursue further scientific investigation.

2

Swallowing a Camel

The Fundamental Argument against Evolution, Part 1

When Charles Darwin tried to explain how his theory of evolution could account for complex organs such as the eye, he began by admitting that the whole idea seemed absurd. He then tried to explain how it could be possible that such organs could arise on their own, but his reasoning was not very convincing. This is not a slight against Darwin. The great naturalist put forth the best argument possible. It's not that Darwin didn't do a very good job; it's that the whole idea was—well, absurd.

The most complex thing we know of is life. From bacteria to elephants, living organisms are by far the most amazing, intricate machines in the world. They inspire awe that even the normally dry, staid textbooks cannot hold back. The last thing that comes to mind is that life somehow arose on its own, spontaneously forming as a result of natural forces.

Living things don't look like they evolved. This is the fundamental argument against evolution, and it is obvious to everyone. One need not be a biologist to know that Darwinism is a stretch. It is asking us to believe in something that seems highly improbable.

A Parable for Evolution

Consider a simple analogy. What if a spacecraft from a distant civilization comes to planet Earth and dazzles a crowd of people with its amazing maneuvers? As it speeds across the heavens, the crowd looks on in disbelief. It is so advanced it can even produce new spacecraft when needed. And when the spacecraft lands, the people gather around to get a closer look at the awesome ship. They wonder where it came from and how it was constructed. Certainly it must be from a very advanced civilization. But then one man wonders if perhaps it had arisen spontaneously—a result of the natural laws of the universe. The rest of the crowd makes no response, but their shuffling feet and sideways glances give away their suspicion that the man is not of sound mind.

Analogies are not perfect. Could it be that this one misrepresents the situation? Could there be something special about carbon-based life-forms that makes them different from inorganic machines such as the spacecraft? Can highly complex organic life spontaneously arise? Well, perhaps so, but we haven't discovered such a magical property yet.

If there is any doubt about the complexity of organic life-forms, then we need look no further than a single-celled bacterium. A single-celled organism contains processes so intricate and complex that the idea that they came about on their own is simply bizarre.

Consider the basic workings of the cell. The information for running the cell is stored in the DNA. To access the information, part of the DNA is copied. It is then used, for example, to construct a protein. The choice of which type of protein to construct is critical. Proteins are highly specific—they do one, and only one, job. If you just ate a big meal, then you'll need some digestive enzymes. If you get a burn, then you'll need some heat shock proteins. The information in the DNA is accessed as it is needed.

This may sound straightforward and even simple, but it is not. In Darwin's day not too much was known about the inner workings of the living cell. He speculated about how these workings may be rather simple. Could it be that the biological world we see around us is all built up from simple building blocks? Might the astronomical number of cells and their many molecules add up to a complex world by virtue of their quantity rather than their quality? Any notions like these that may have been held in the nineteenth century were quickly dispelled by the twentieth century's profound discoveries in molecular biology. Let's have a look at just a few of those discoveries.

DNA Stores Information

DNA is a double helix. It contains two strands that are twisted about each other, just as a rope might consist of two smaller ropes wrapped around each other. Both DNA strands consist of a series of molecules glued together in sequence. The molecules are called nucleotides, and if you stretched out the strand, the nucleotides would line up in a row, like beads on a string. These nucleotides are precisely the DNA's information. If each nucleotide represents a letter, then a stretch of DNA is a sentence. DNA strands use four different kinds of nucleotides, so in this language there are only four letters.

A tremendous amount of effort has been spent on understanding how these letters are read. How is the information stored in the DNA accessed? In 1953, before the DNA structure was deciphered using X-ray photographs, the great American chemist Linus Pauling predicted that the nucleotides pointed outward from a helix. A few years earlier Pauling had successfully predicted the helical structure of proteins. Now he would try to do the same for the DNA structure. If the nucleotides pointed outward, Pauling reasoned, they could be read without having to pull apart the DNA. But this time Pauling's intuition failed him.

Not long after that, Francis Crick and James Watson, working in Cambridge, England, solved the puzzle. Using Rosalind Franklin's X-ray measurements, they determined not only that DNA normally is a double helix, but that the nucleotides are pointed inward, toward the center of the helix. "We have discovered the secret of life," proclaimed Crick at a nearby Cambridge pub.

It seems often the case that scientists overestimate the state of science. In the eighteenth century, with Newtonian physics firmly in place, the great French scientist Pierre Laplace proclaimed that science could, in principle, predict the future. And not long before the twentieth century's breakthroughs in relativity and quantum mechanics, there were those who felt physics had nothing more to discover. All that was left was to fill in the details. Such high confidence in the state of science had to give way to the hard facts. Science will probably never be able to predict the future, and the field of physics is alive and well, full of unanswered questions.

Likewise, Watson and Crick's new DNA model was a great break-through, but it is not *the* secret of life. It probably raised more questions than it has answered. How, for example, was the DNA information located and read if the nucleotides were tucked away in the interior of the helix? And how was the information later interpreted? These questions have since been largely answered (and others are taking their place), and the picture that is coming together appears to be anything but the product of evolution.

Where to Copy a DNA Strand

Consider how the DNA information is accessed and used in *E. coli*, a common bacterium. A large protein complex, about twenty times the size of an average protein, attaches itself to the DNA. The complex will ultimately construct a copy of one of the DNA strands. The copy, or transcript, is chemically slightly different and is called RNA. The protein complex is called RNA polymerase because it constructs an RNA strand, or polymer.

There are about 7,000 RNA polymerases in an *E. coli* cell, and thousands of them are making RNA transcripts at any given time. But which segments of DNA should the RNA polymerase copy? Just as a book is organized into a series of chapters, the DNA information is organized into a series of genes. The information in a gene is used to construct big molecules, such as proteins. How does the RNA polymerase know which protein is needed and where on the DNA to start copying?

The RNA polymerase needs to be able to search rapidly, because the starting locations of all the genes make up something like 0.001 percent of the total DNA—finding them could be like finding a needle in a haystack. But once the right starting location is found, the RNA polymerase needs to lock on tightly so it can copy the DNA. Furthermore, when not busy copying, the RNA polymerase waits patiently by attaching itself randomly to DNA. The cell's complement of inactive RNA polymerase is not concentrated in one place, but is stored randomly along the DNA.

The RNA polymerase accomplishes these disparate tasks with the help of a small helper molecule called the sigma factor. With sigma attached, the RNA polymerase loses its affinity for DNA in general but gains a strong affinity for the appropriate starting location. Hence, RNA polymerase detaches from its resting place, somehow searches along the DNA, and reattaches when it finds the starting location.

Once the copying process is successfully under way, the sigma factor is released so the RNA polymerase loses its strong affinity for the starting location. It is now free to move along the DNA and construct the RNA transcript. The sigma factor is like a switch that changes the RNA polymerase function, and it comes in a variety of forms that help RNA polymerase to locate different genes.

Just as sigma alters the RNA polymerase, so too does RNA polymerase alter sigma. The sigma factor attaches to the DNA starting location when it is bound to the RNA polymerase, but the sigma factor by itself does not attach to the DNA. If the sigma factor did attach to DNA, it would prevent its attachment to the RNA polymerase. This would also obscure the DNA starting locations. There is more to learn about this process, but we can already see that it is complicated and intricate.

How to Copy a DNA Strand

Once the RNA polymerase finds the starting location, it must open up the DNA double helix in order to "read" the nucleotides within and make its copy. For although it was possible to find the starting location without opening the DNA, the copying process requires an open DNA strand to work from.

How does the RNA polymerase open the DNA double helix? One way to make the job easier would be to unwind the helix a bit. Imagine that you grabbed a rope in two different places with your hands. If you twisted your hands in opposite directions so as to unwind the rope, the different strands in the rope would tend to come apart.

Something like this happens in the cell as two different proteins, in front of and behind the RNA polymerase, apply opposite twist to the DNA strand. With the helix unwound, the RNA polymerase moves along the strand, copying one nucleotide at a time. In this process, nucleotides in the environment are used to build an ever-growing RNA strand that is a duplicate of one of the DNA strands.

Twenty to fifty nucleotides are added to the RNA strand per second, with an error rate of about 1 wrong nucleotide per 10,000. This means that something like one gene in ten will have one wrong nucleotide. This error rate is tolerable for several reasons. First, when a given protein is needed, many transcripts may be constructed in order to produce large quantities of the protein. An error in one transcript will not likely be repeated in the next. Second, there are two built-in safety buffers in the construction and assembly of proteins. As we shall see, a single wrong nucleotide will not likely affect the protein function.

Controlling the Copy Machine

How does the RNA polymerase know which gene to make copies of at any given time? The sigma factor helps it to find the starting location and can, to a certain extent, help select the right gene. But there are more than 4,000 genes in E. coli. How does the RNA polymerase, even with sigma's help, find the right one?

We cannot yet answer this question completely, but the basic idea is now well understood. In 1961 two French scientists, Jacques Monod and François Jacob, proposed a revolutionary idea: that genes could be turned on and off by other genes. The idea turned out to be a good one. Simply put, they said that most genes make proteins to service the cell, but the production of those proteins can be controlled by other genes. In other words, some gene products do the work of the cell, such as digesting nutrients, but others regulate that work.

For example, five different proteins are required to manufacture the amino acid tryptophan. Since E. coli usually needs tryptophan, RNA polymerase routinely makes copies of the genes for these proteins. But sometimes tryptophan is freely available in the environment. In this case there is no point to manufacturing it, and RNA polymerase is blocked by a regulatory protein called tryptophan repressor.

Tryptophan repressor is always hanging around, but on its surface there are two sites that tryptophan itself fits into rather nicely. When tryptophan is freely available, then some of it docks with the tryptophan repressor, causing the repressor to change shape. The result is that tryptophan repressor now binds to the DNA at just the right place to block RNA polymerase.

The tryptophan repressor protein regulates the production of tryptophan, and the switch gets thrown by tryptophan itself. When there is no free tryptophan the tryptophan repressor is turned off, but when tryptophan is available the tryptophan repressor is turned on. This is called negative control, because the tryptophan repressor turns off the copying process when bound to the DNA.

A slightly more complicated example involves the digestion of the milk sugar lactose. Lactose consists of the two simple sugars, glucose and galactose glued together, and it is digested by three different proteins.

The E. coli cell does not want to produce the three proteins if (1) lactose is not present in the environment or (2) simple glucose, a preferred nutrient, is available. How can E. coli produce the proteins only when lactose *is* available and glucose *is not* available? In this case there must be two regulatory proteins involved. The presence of lactose *turns off a repressor protein*, and the absence of glucose *turns on an activator protein*. These two

19

proteins exercise negative and positive control, respectively. In order to copy the three genes, the repressor protein must be detached, but the activator protein must be bound to the DNA.

Notice that all these regulatory proteins must have at least two binding sites, one for the switching substance (tryptophan in the case of the tryptophan repressor) and one for the DNA. And the former must control the latter. That is, the DNA binding function must be dependent on the other site.

The regulation of tryptophan synthesis and the regulation of lactose digestion are classic textbook examples, and things quickly become more complicated in other cases. Even these simple examples are more complicated than this brief sketch suggests. Nonetheless, it should be clear that controlling the copying action of RNA polymerase is an exquisite process requiring many details to be worked out in advance. It is not the sort of thing that lends well to Darwin's idea of unguided evolution. Not surprisingly, evolutionists have little more than speculation to explain how such systems arose.

For example, one text makes the claim it is "easy to see" how negative control might evolve. The reader is told that the regulation by negative control turns off proteins when they are not needed. This makes the cell more efficient, and therefore this would be selected by evolution.[1] This much, of course, is obvious. A better design is more efficient, so natural selection would choose it. But the text completely ignores the problem of how such a complex system could possibly arise on its own. Darwin's idea is that evolution selects from *naturally* occurring variation, but how does such a phenomenally complex system just arise on its own so that it can be selected?

But even this type of credulous thinking runs into problems. Even when evolution is granted every advantage and details are conveniently ignored, biology still makes it appear questionable. Take, for example, the case of positive control. Here the gene is not copied without the aid of the regulating protein and switch molecule. Did evolution create a dormant gene and then add regulation? Not likely, for this would suggest that evolution has a forward planning capability.

The other alternative is that evolution began with a working gene that simultaneously became both dormant and regulated. This would require the introduction of the regulation apparatus, including a functional activator protein and switch molecule. Perhaps the regulation apparatus was conveniently in place already and needed only to be modified. Thankfully, this occurred at those genes where positive control was needed. Our textbook author writes: "It is less obvious how positive control evolved, since the cell must have had the ability to express the regulated genes even before any control existed. Presumably some component of the control system must have changed its role. Perhaps originally it was used as a regular part of the apparatus for gene expression; then later it became restricted to act only in a particular system or systems."[2] The speculation is so rampant that evolution looks like a "just-so" story.

Building a Protein

The RNA transcript created by RNA polymerase typically serves as the instructions for making a protein. This is done by feeding the transcript through a translation machine called a ribosome. In World War II the Germans used a famous cipher machine called Enigma to encode and decode their military communications. If you fed the Enigma machine a coded message, it would decode the message and produce the readable form.

Similarly, the ribosome decodes the transcript that RNA polymerase produces. The transcript is an RNA strand consisting of hundreds of nucleotides lined up like molecular letters. There are four different letters (A, T, G, and C), and they form words, each of which are three letters. For example, a word might be GCC or TAC. There are a total of sixty-four different words. Three of the words serve as a stop signal, and the other sixty-one code for a particular amino acid.

Using this code, the ribosome translates a string of nucleotides into a string of amino acids and terminates when it hits a stop signal. The coded message that is fed to this molecular translating machine is the RNA strand; the decoded message that is produced is a chain of amino acids that will become a protein.

The sixty-one different words code for only twenty different amino acids. This means that many of the words code for the same amino acid. In other words, a given amino acid may be represented by several words. Therefore, errors in the RNA strand or in the translation process may end up to be inconsequential. For example; the words AAA and AAG both code for the amino acid lysine. So if AAA was the intended message, but AAG was read, the decoded message would not change.

The code is sometimes referred to as the universal genetic code, because it is used in the cells of practically all species. Everything from whales to oak trees uses the same genetic code.

How does the ribosome translate the RNA strand into a chain of amino acids? We still do not fully understand the process, but we do know that the ribosome uses a small army of "reader" molecules. The reader molecules have on one end three nucleotides that recognize specific words on the RNA strand. On the other end the reader molecules have an amino acid attached.

The reader looks for particular words on the RNA strand that correspond to its amino acid, according to the DNA code. As the ribosome works its way through the RNA message, reader molecules are brought in one at a time that match up with the encoded words. The ribosome takes the amino acids from the readers and attaches them to the growing chain of amino acids, sometimes as fast as twenty per second.

The "stagehands" who make this incredible dance possible are a class of proteins that attach the amino acids to the reader molecules. These proteins are the real implementers of the genetic code—they attach the precise amino acid that is coded for by the word that will be recognized by that reader molecule.

For example, one of these proteins will attach lysine to the reader molecule that will recognize the AAA word. Those reader molecules that recognize AAA in the RNA strand will present lysine to the ribosome as the next amino acid, just as the DNA code requires.

Of course, in this brief description I've left out many details. For example, there are at least a couple of "proofreading" mechanisms that help the process attain its high accuracy (on average, several thousands of amino acids translated will have only a single error). But even this brief sketch should make it clear that the cell's construction of proteins seems to be "inexplicably complex," as one text put it. "The complexity of a process with so many interacting components," the text continues, "has made many biologists despair of even understanding the pathway by which protein synthesis evolved."[3]

The challenge for evolutionists is to find a step-by-step buildup of the process. At each step the process must be functional. Also, an often-overlooked requirement is that each step must have some reasonable likelihood of randomly occurring in a reasonably sized population of individuals.

Darwinists are not short on speculation about how the DNA code and protein synthesis could have evolved. But their speculation fails to meet these criteria. They do not show how each step is functional or likely to occur. In fact, they do not even lay out the step-by-step buildup. Instead, we are given vague explanations about how evolution could have worked.

One idea, for example, is that the cell's inner workings did not at first depend on proteins. Instead, RNA molecules both stored information and carried out the actions necessary to sustain the cell. This idea has not been developed with any sort of detail to make it believable, but evolutionists now routinely refer to it as though it is a more or less likely explanation. And so our text confidently concludes that the great complexity that brought evolutionists to despair is now resolved as they have a new way of viewing the pathway by which protein synthesis evolved.

Folding a Protein

Once the chain of amino acids is complete, it folds up to form a functional protein. These are called globular proteins, and their compact shape is very important. Each protein folds up into the same shape practically every time. Proteins do their job by docking with other molecules, and the shape encourages the docking to take place. Without the right shape the protein would not function correctly.

Not only are proteins good at folding into the same shape every time, they also fold quickly, usually in just a fraction of a second. Proteins need to fold fast and consistently in order for the cell to work well. But proteins also need to be structurally flexible. A rigid protein would not be forgiving in the docking process, and proteins often change their shape as part of doing their job.

Also, proteins need to be quickly disassembled once they are no longer needed, and rigid proteins would be more difficult to take apart. Thus, proteins need to be flexible, and this means they must be marginally stable, just barely maintaining their shape.

It seems like a paradox for a molecule to form quickly and consistently, yet be on the verge of coming apart. It would seem that a molecule would form quickly because there are strong forces encouraging it to form. How can the molecule then come apart so easily?

Questions like these about protein folding have been studied since the 1960s, yet many remain unanswered. One goal of the research is to be able to predict the protein shape given a new sequence of amino acids. This problem is like trying to break a code. Typically, the amino acid sequence determines the structure. It can be thought of as a message that describes the structure, and the translation from message to structure is done according to the laws of physics and chemistry.

Not knowing the protein folding code makes it more difficult to make sense of the huge volumes of genome data that are now being read from the DNA of many species. Most of the interesting parts of the genome code for proteins, and knowing the shapes of those proteins would help us understand the genome as a whole.

Though we don't completely understand protein folding, we have nonetheless learned much about the problem. One thing that is becoming clear is that very few amino acid sequences are functional. You'd need to look through many billions of different sequences before finding one that is functional. In other words, finding a sequence of amino acids that provides a useful function is like the proverbial needle in a haystack. If evolution started searching for a functional protein sequence when the universe began, and it looked at one sequence per picosecond (there are a million million picoseconds in a second), it still would not likely have succeeded, even for a relatively small protein.

For evolution to construct a protein gradually over time it would require a long series of slightly different functional sequences leading to the protein. Since we don't find evidence for this, evolutionists must believe that proteins are somehow created by unlikely mechanisms that only rarely can stumble on a functional sequence.

The Conditions for Complexity

It is ironic that evolutionists sometimes point to protein folding as an example of complexity spontaneously arising. It is true that protein folding is an example of this, but only because the conditions are just right for it to happen. For example, you could make the word *evolution* appear spontaneously on your computer screen, but it would require, behind the scene, a carefully written computer program to make the word appear. Likewise, a protein consistently folds because a very particular sequence of amino acids has been connected together in a hospitable environment.

Protein folding is just one of the many examples in biology of complexity spontaneously arising because the conditions are just right. Organisms and populations respond to changing conditions in a variety of amazing ways. Bacteria create new proteins and insects develop resistance to pesticides. It may appear that complexity is arising spontaneously, but behind the scenes there is a clever adaptation machine at work.

Darwinists claim that such adaptations are powerful evidence for their theory. Are new proteins and pesticide resistance not examples of evolution in progress? The problem is these adaptations are produced by a machine that appears to be set up to produce such changes. Rather than mutations aimlessly exploring new designs, we are apparently witnessing the actions of a complex and robust machine.

For example, when a population of bacteria is subjected to harsh conditions, they tend to increase their mutation rate. It is as though a signal has been sent saying, "It is time to adapt." Also, a small fraction of the population increases their mutation rates even higher yet. These hypermutators ensure that an even greater variety of adaptive changes are explored. They may be essential to the survival of the bacteria population in some cases. Furthermore, the mutations themselves do not appear randomly throughout the genome but are concentrated in certain areas that can produce helpful changes. In other words, pathways of adaptation are, to a certain extent, already laid out.

Likewise, our knowledge of pesticide resistance in insects suggests a similar scenario. One study of the common fruit fly found that pesticide resistance arises from a gene that has been present all along. The gene serves to break down the pesticide. It used to be less active, but now it is more active in resistant flies. A special signal was inserted into the gene to lift production constraints. It appears that pesticide resistance is conferred by flipping a switch on the genetic production line rather than creating a new factory.

These findings are awkward for evolution. Instead of single mutations leading to a new functionality one step at a time, we must believe that evolution produced this marvelous machine by which more complicated changes can occur. Genomes, Darwinists must say, have evolved the capacity to respond to predictable environmental challenges with predictable changes. Mutational rates are sped up when needed and concentrated in those locations where needed, and fortunately multiple mutations can occur in a single step.

So evolution has produced a machine that can evolve. Evolution, Darwin's general principle, has created evolution, the biological function. This account is hardly compelling, for it has too much serendipity and too little accountability. Adaptations in biological organisms such as bacteria and insects do not serve as good evidence for their evolution.

3

Swallowing a Camel

The Fundamental Argument against Evolution, Part 2

The cellular processes that use the DNA information to construct proteins are phenomenally complex. In this chapter we continue the story with a few of the many examples of complexity at higher levels.

Proteins in Action

Once folded, proteins go about their work, which can be one of many different jobs. Proteins are the workers behind just about every task in the cell, including digestion, chemical synthesis, energy conversion, cell reproduction, and, as we have seen, making new proteins. And like a finely tuned machine, proteins do their work very well.

Enzymes are a good example of how well proteins can work. Enzymes are proteins that accelerate biochemical reactions that must take place in the cell. There are literally thousands of chemical reactions taking place simultaneously in the cell at any given time. But these chemical reactions are usually too slow. They need to be sped up if they are to keep up with the busy cell.

We cannot speak of a typical chemical reaction time, as times vary from minutes to millions of years. For the cell it is all the same—a minute may as well be a million years, for it needs these reactions to happen in but a tiny fraction of a second. This is the biologically relevant time period, and for every one of these required biochemical reactions there is an enzyme that speeds it up so that it is biologically relevant.

This is one of the amazing things about enzymes. Whether a reaction takes one minute or one million years, its enzyme raises the rate to the biologically relevant level. Chemists speak of catalytic power—the ability of an agent to accelerate a reaction—and the catalytic power of many enzymes is enormous, far greater than industrial catalysts. Enzymes achieve these results without the benefits of high temperature, pressure, chemical concentration, or other extremes.

There is, for example, the incredibly proficient enzyme OMP decarboxylase, which helps construct nucleotides. The reaction it catalyzes would normally require about 78 million years, but with OMP decarboxylase the reaction takes place in a fraction of a second. And for sheer speed there is the enzyme catalase, which can break down 40 million molecules of hydrogen peroxide (H_2O_2) per second.

Enzymes carry out their duties by performing very specific, detailed chemical reactions. Chemical bonds are formed and broken, electrons are shuffled, protons are donated, etc., as the substrate molecules are manipulated.

A classic textbook example is the serine proteases, which snip other protein molecules apart. Typical serine proteases have 245 amino acids. Once they fold up, there is a pocket or depression on the surface, called the *active site*, where the target protein docks and is cut in two.

When you fold a shirt the collar may end up touching the bottom of the shirt. Likewise, when an enzyme folds up, its amino acids from different places may end up close together. In the serine proteases the particular amino acids at the active site are carefully chosen. Histidine-57 (the fifty-seventh amino acid in the chain), aspartate-102, and serine-195 are all consistently found at the active site in serine proteases. This is because the particular characteristics and positions of these amino acids in the active site are responsible for the enzymatic action. They carry out a concerted sequence of actions that snips the target protein.

How does the serine protease know where to make the cut? Different serine proteases have different types of active site pockets. One is rather large and is negatively charged, so it attracts the large, positively charged amino acids arginine and lysine. Another has a medium-sized depression that is oily, so it attracts medium-sized oily amino acids. Yet another has a small depression, so it makes the cut only where small amino acids are located.

The folding of a protein is critical. It determines the final shape and which amino acids go where. Slight changes in the shape of the active site make for a different function. The amino acid sequence determines how it folds up, and it also determines which amino acids will be available at the active site. In other words, the sequence of amino acids has both a structural and a chemical role. It not only must provide the correct shape, but must also provide the correct chemistry.

It is yet another mystery for evolution to determine. How did an unguided process come up with something so exquisite and precise? Again, evolutionists can only speculate. For enzymes that are highly similar, such as the serine proteases described above, evolutionists typically hypothesize that the family evolved from one of them. Given the existence of one serine protease, the evolution of the others is seen as a much easier task.

One problem with this explanation is that it ignores the fact that the various events within the cell are ultimately all connected. The cell is holistic and shows no signs of being the result of a gradual development process. For example, a new enzyme would not evolve if there was no need for it, but if there was a need for it, then the cell would have to be already fairly complex. Is there an evolutionary development path where enzymes can be added here and there? We don't know, because the cell is too complicated.

Perhaps the more pressing problem for evolution is the creation of the first enzyme of the family. How did the first serine protease come about? The probability of a serine protease arising in one piece from a collection of mutations is so remote that even evolutionists agree it is impossible. The development path, say evolutionists, is one of slow increase of complexity and function. But the complexity of the problem is such that we must settle for vague explanations. The evolution of an enzyme is as much a matter of faith as anything else.

Enzyme Pathways and Cycles

Different types of enzymes are often teamed up to carry out a sequence of reactions. For example, in the glycolysis pathway, which occurs in practically all cells, about a dozen enzymes break down the six-carbon sugar known as glucose into two three-carbon molecules. Like a factory production line, each enzyme catalyzes a specific reaction, using the product of the upstream enzyme and passing the result to the downstream enzyme. If just one of the enzymes is not present or otherwise not functioning, then the entire process fails to work.

In addition to breaking down glucose, glycolysis also produces energy-carrying molecules called ATP. These are in constant demand in the cell, as they are used wherever energy is needed. Like most pathways, glycolysis is interconnected with other pathways within the cell. The molecular products of glycolysis are used elsewhere, and so the rate at which the glycolysis pathway proceeds is important. Too fast, and its products won't be useful; too slow, and other pathways have to slow down.

Glycolysis is regulated in a number of ways. We saw earlier that the process for copying genes is elegantly controlled. For example, when a protein is not needed, the copying of its gene is temporarily shut down. Likewise, pathways are precisely controlled. The first enzyme in the glycolysis pathway is regulated by its own product. This enzyme alters glucose to form an intermediate product, but if the rest of the pathway is not keeping up, then the intermediate product will build up, and this will cause the enzyme to shut down temporarily. The enzyme is designed to be controlled by the presence of its product.

Two other enzymes in the pathway have even more sophisticated regulation. They are sensitive to a number of different molecules that either increase or decrease the enzyme activity. For example, these enzymes are partly controlled by the energy level of the cell. This makes sense, since glycolysis helps supply energy to the cell. A good indicator of the cell's energy level is the relative concentrations of ATP and spent ATP. High levels of ATP indicate a strong energy supply. Hence, the enzyme activity is inhibited (and therefore the glycolysis pathway is slowed) when ATP is abundant. But high levels of spent ATP counteract this effect.

How do these molecules control enzyme activity? The molecules are tiny compared to the big enzymes they control. Just as a small key is used to start up and turn off a big truck, so too these small molecules have big effects on their target enzyme. And just as the truck has an ignition lock that can be turned only by the right key, so too the enzyme has several docking sites that are just right for a particular small molecule, such as ATP.

Not only does ATP fit just right into its docking site, but it perturbs the enzyme structure in just the right way so as to diminish the enzyme activity. There is another docking site that only a spent ATP will fit into. And if this occurs, then the enzyme structure is again perturbed just right, so as to encourage activity and reverse the ATP docking effect.

The details of enzyme regulation are a topic of current research, and many different modes of regulation have been discovered. Suffice it to say that the regulation depends on precise interactions between the enzyme and regulating molecule. As we saw above, the enzyme amino acid sequence determines its structure and chemical properties along the structure. These are critical for the regulation to succeed.

The enzyme must fold up into just the right shape so that docking sites are available. The regulating molecules must not only fit into those sites, but induce the appropriate effect in the enzyme in order to control the activity.

How are we to imagine the evolution of metabolic pathways, with their specialized enzymes and precise control? Evolutionists have been working hard to explain their evolution, but not surprisingly they are unable to give any sort of detailed or specific account. One recent paper, for example, explains the evolution of a complicated pathway as an example of evolutionary opportunism.[1] Most of the enzymes, so the argument goes, were already available because they would have served other purposes. Only one brand-new enzyme would have been required.

Reducing the problem to the creation of only a single enzyme sounds good, but we must remember that even this is an enormous problem. There must be a need for the enzyme, though the cell must be able to get by without it. Then random mutations must stumble upon an amino acid sequence that results in that enzyme. Of course the enzyme must be produced and be available where needed. Though the newly minted enzyme was not previously part of the pathway, it now must join in at the right place without upsetting the production flow.

Furthermore, the problem has not really been reduced to a single enzyme. Even if the other enzymes already existed, they of course must have evolved as well. None of these difficulties seem to daunt evolutionists as they make speculative proposals for how their unguided process creates the wonders of the world.

Nerve Cells

There are a great many more proteins that do a variety of jobs. Some are attached to the flexible membrane that encapsulates the cell. A fascinating example of membrane proteins is found in nerve cells.

Nerve cells have a long tail, which carries an electronic impulse. The tail can be several feet long, and its signal might stimulate a muscle to action, control a gland, or report a sensation to the brain.

Like a cable containing thousands of different telephone wires, nerve cells are often bundled together to form a nerve. Early researchers considered that perhaps the electronic impulse traveled along the nerve cell tail like electricity in a wire. But they soon realized that the signal in nerve cells is too weak to travel very far. The nerve cell would need to boost the signal along the way for it to travel along the tail.

After years of research it was discovered that the signal is boosted by membrane proteins. First, there is a membrane protein that simultaneously pumps potassium ions into the cell and sodium ions out of the cell. This sets up a chemical gradient across the membrane. There is more potassium inside the cell than outside, and there is more sodium outside than inside. Also, there are more negatively charged ions inside the cell, so there is a voltage drop (50–100 millivolts) across the membrane.

In addition to the sodium-potassium pump, there are sodium channels and potassium channels. These membrane proteins allow sodium and potassium, respectively, to pass through the membrane. They are normally closed, but when the electronic impulse travels along the nerve cell tail, it causes the sodium channels to quickly open. Sodium ions outside the cell then come streaming into the cell down the electrochemical gradient. As a result, the voltage drop is reversed and the decaying electronic impulse, which caused the sodium channels to open, is boosted as it continues on its way along the nerve cell tail.

When the voltage goes from negative to positive inside the cell, the sodium channels slowly close and the potassium channels open. Hence, the sodium channels are open only momentarily, and now with the potassium channels open, the potassium ions concentrated inside the cell come streaming out down their electrochemical gradient. As a result, the original voltage drop is reestablished.

This process repeats itself along the length of the nerve cell until the impulse finally reaches the end of the cell. Although I've left out many details, it should be obvious that the process depends on the intricate workings of the three membrane proteins. The sodium-potassium pump helps set up the electrochemical gradient, the electronic impulse is strong enough to activate the sodium channel, and then the sodium and potassium channels open and close with precise timing.

How, for example, are the channels designed to be ion-selective? Sodium is about 40 percent smaller than potassium, so the sodium channel can exclude potassium if it is just big enough for sodium. Random mutations must have struck on an amino acid sequence that would fold up just right to provide the right channel size.

The potassium channel, on the other hand, is large enough for both potassium and sodium, yet it is highly efficient. It somehow excludes sodium almost perfectly (the potassium-to-sodium ratio is about 10,000), yet allows potassium to pass through almost as if there were nothing in the way. The solution seems to be in the particular amino acids that line the channel and their precise orientation. For potassium, moving through the channel is as easy as moving through water, but sodium rattles around—it fits in the channel, but it makes less favorable interactions with the amino acids.

Feedback Signals

Nerve cells are constantly firing off in your body. They control your eyes as you read these words, and they send back the images you see on this page to your brain. They, along with chemical signals, control a multitude of processes in our bodies.

For example, our cardiovascular system runs twenty-four hours a day, seven days a week without our giving it a conscious thought. Our nerves control muscle motion that expands our lungs to draw in outside air and pump blood through the heart.

The blood leaves the heart carrying oxygen to the various parts of the body. Oxygen is important for making the energy molecule ATP. Glycolysis uses no oxygen and makes only two ATP molecules. Other pathways that use oxygen make almost twenty times more ATP.

This more prolific ATP manufacturing process occurs in a machine within the cell called the mitochondria. The mitochondria's method for constructing ATP is quite clever, and it took many years of research to uncover its inner workings.

In mitochondria the oxygen delivered by the blood is not used directly to construct ATP. Instead, the oxygen is used with an elegant series of reactions to set up a proton gradient across a mitochondrion's inner membrane. As a battery forces electricity to flow through a wire, so the proton gradient forces the protons to flow back across the membrane.

There is a membrane protein called ATP synthase that has a channel for the protons to flow through. Like a waterfall turning a generator, the protons turn a crank in the ATP synthase. It appears that the crank is not straight but has a curve in it. As it rotates, it apparently pushes on other parts of the ATP synthase protein, causing the structure to change shape.

As the protons continue to flow, the ATP synthase goes back and forth between different conformations. The ATP synthase structure and its conformational changes are just what is needed to capture spent ATP molecules and recharge them. Just as a hydroelectric dam converts water pressure to electricity, the ATP synthase converts proton pressure to chemical energy, in the form of the ATP molecule.

After the blood donates its oxygen for this cause, it heads back to the heart, carrying carbon dioxide, a waste product of ATP construction. The blood is drawn in by the heart and pumped along to the lungs, where it is exposed to inhaled air across a gas-permeable membrane. Here the blood and air trade oxygen for carbon dioxide, and the replenished blood is drawn back to the heart to begin another cycle.

The star player in this molecular choreography is the hemoglobin molecule. Hemoglobin is a complex of four similar proteins attached together. When the blood passes by the lungs, it picks up oxygen molecules from the air. But the blood can hold only so much oxygen—while it gains oxygen atoms *from* the air it also loses oxygen atoms *to* the air. The blood alone has no way of holding onto the oxygen. Hemoglobin solves this problem. It binds the oxygen, so it cannot escape back to the air. The result is that the blood can carry much more oxygen.

When the blood then moves on to the various parts of the body, the hemoglobin molecules unload the oxygen where needed. But time is of the essence, and as hemoglobin loads and unloads oxygen it must do so quickly. Consider a dump truck carrying a load of dirt. The truck would be practically worthless if it couldn't tilt up its box so as to quickly dump the dirt. Likewise, as the hemoglobin passes by the lung there is little time to load its complement of oxygen, and as it passes by an oxygen-starved muscle there is little time to unload the oxygen.

Hemoglobin speeds things up via minor changes in its structure. Earlier we saw that enzyme regulation can come about by slight structural changes. Likewise, conformational changes in this amazing protein can make hemoglobin a quick oxygen loader or a quick oxygen unloader.

There are many more examples of complicated, efficient, and robust mechanisms in biology. There is the thermostat that is built into our brains. It controls blood flow in response to body temperature, sending more to the skin surface when hot and diverting it away from the skin when cold. It also activates sweat glands when hot and muscle shivering when cold. There is the blood sugar sensor in the pancreas. When blood sugar levels are low, the pancreas releases glucagon, a hormone that instructs the liver to release stored glucose. But after a big meal, the pancreas releases insulin, which instructs the liver to store the abundant glucose.

Biology is full of incredibly elaborate, complex machines. If you are beginning to suspect that Darwinism has no compelling explanation for them, you're right. Aside from vague hypotheses that have more speculation than hard fact, evolutionists have no idea how such machines could have come about by the unguided forces of nature.

When I was a boy I read Hans Christian Andersen's classic tale "The Emperor's New Clothes." The story is about an emperor who is fooled into wearing no clothes and the mob mentality that overtakes his subjects as they too are led to go along with the charade. All the people in the kingdom are told that the emperor has beautiful new clothes, and the emperor is convinced as well. When the people see the smiling emperor with no clothes, no one wants to point out the obvious.

When I first read the story I was unimpressed. What was the point? The story certainly had no bearing on the real events of the world. Such an obviously false and absurd charade could never actually take place, and if it did, large numbers of people would never go along with it.

But now I appreciate Andersen's tale. It is indeed possible for people to go along with bizarre explanations. The problems with evolution are evident in nature itself. Biology is full of amazing designs whose evolution would apparently constitute nothing less than a miracle.

The Problem of Complexity

Darwin knew well enough that evolution would seem absurd and that he would have to supply a good reason for the hope he had in natural forces. Of course there is no scientific argument for how the most complex thing we know of arose on its own.

It is simple intuition that good things don't just happen on their own. The ground does not produce a useful harvest without painful toil. A dining room does not organize and clean itself after a feast. And nothing, from clothing to cars, self-assembles.

How could Darwin convince the world that evolution could create complexity? He had no strong scientific explanation, so he shifted the burden of proof. Rather than requiring evidence showing that evolution could create complexity, Darwin suggested that there was no *counterevidence*. He allowed that if the skeptic could find a complex organ that evolution *could not* produce, then the theory would be disproven. But it would be impossible for a skeptic to *prove* that evolution could never create complexity, for that would be tantamount to proving a universal negative.

Darwin made things easy on his theory by inverting the question. Rather than asking the question "How much positive evidence is there that complexity can arise on its own?" he asked "Is there negative evidence to disprove the idea?"

There is, of course, an abundance of negative evidence. Our fundamental understanding of the world informs us that the spontaneous evolution of highly complex, intricate structures is unlikely. But to *disprove* the idea is far more difficult. There are plenty of unlikely ideas that nonetheless cannot be absolutely disproven.

Darwin's argument was not in the scientific spirit, for one does not propose an unlikely and unproven theory and justify it because it cannot be disproven. It was the best argument available to Darwin, and he used it skillfully. With very little empirical evidence, Darwin was able to remove a major problem for his theory.

After Darwin, evolutionists rarely needed to defend the theory against the problem of complexity. Complexity went from being a liability to a research program. When critics raise the complexity problem, they are told that the argument is faulty because it is from scientific ignorance. What is needed is more research. Evolution, they say, may seem unlikely, but we cannot say it is improbable because we do not fully understand natural processes. Instead of requiring scientific theories to be likely, evolutionists require merely that they are naturalistic. In the meantime we must settle for vague and cursory explanations. This or that example of complexity evolved because that is the way evolution occurred.

The fields of mathematics and logic can provide objective proofs for their results, but science is ultimately subjective. Whether the arguments and evidence for a theory are worthy is ultimately a matter of judgment. This can be a strength, as it allows scientists to imagine and hypothesize in the absence of solid proof. But this can also be a great weakness, as it allows for fine-sounding arguments that have little empirical support to slip past the gates of science.

4

Straining at the Gnat

The Scientific Evidence for Evolution, Part 1

Living organisms are incredible machines. They inspire awe and wonder. Charles Darwin's idea that life arose on its own is not what comes to mind when studying life. Evolution is not intuitive, but this does not mean it is wrong. Intuition is not always an accurate guide, and when scientific data contradict our intuition, we should follow the data. Darwin's close friend and ally Thomas Huxley made this point when arguing for evolution. The scientist should, urged Huxley, always be ready to set aside preconceptions and, like a child, follow the scientific data. I couldn't agree more.

The theory of evolution makes extraordinary claims. Did the most complex things we know of arise by the unguided, natural forces of physics and chemistry? Perhaps, but extraordinary claims require extraordinary evidence. Huxley urged that we follow the data because he felt they supported Darwin's theory of evolution. Let's take Huxley's statement at face value and, like a child without preconceptions, examine the scientific data.

The Fossil Evidence

Evolutionists routinely claim that the fossil evidence proves evolution. Fifty years ago the great evolutionist George G. Simpson wrote that "there really is no point nowadays in continuing to collect and to study fossils simply to determine whether or not evolution is a fact. The question has been decisively answered in the affirmative."[1] Tim Berra claims that the "fossils provide hard evidence that evolution has occurred"[2] and that the fact that "even one transitional fossil is found is a sufficient demonstration of evolution and a resounding falsification of creationism."[3] Kenneth R. Miller claims that the fossil data indicate that "descent with modification, which most of us prefer to call evolution, really happened."[4]

The fossil remains of strange-looking creatures have been observed since antiquity, but it wasn't until the eighteenth century that the fossil record began to be systematically and scientifically analyzed. By Darwin's day it was well known that very different creatures once roamed the earth. Those creatures appeared, persisted, and mysteriously disappeared in the fossil record. There was a progression from lower to higher forms, but it was not gradual or continuous. Why? Were the fossils a spotty recording of natural history, giving only an occasional snapshot of the biological world? Or did the fossils reveal a rather uneven process of persistence and abrupt change?

True to his creed, Huxley urged Darwin to stick to the raw data. Evolution, it seemed to Huxley, ought to be allowed to take jumps, for that is what the fossil record revealed. But Darwin did not heed the advice of his confidant. Evolution, Darwin maintained, must be gradual and the fossil record must be highly imperfect. Darwin was now taking leave of the data, and he knew not everyone would go along. The imperfection of the fossil record was a premise of his theory. Those who disagreed with the premise, Darwin admitted, would rightly reject his entire theory.

Does Nature Take Jumps?

Darwin's strong stand against discontinuities is not surprising, for such sentiment had been growing for many years. A century before Darwin, Anglican bishop Joseph Butler argued that the Bible's miracles were probably true. But in the late eighteenth and early nineteenth centuries miracles were viewed with increasing suspicion. At best they couldn't be confirmed; at worst they violated natural laws, which were increasingly being viewed as the basis of knowledge and progress. By Darwin's day many Anglicans were calling for a purely mechanistic creation. A God who needed miracles was not much of a craftsman and not very worthy of worship.

This sentiment easily jumped species when early scientists called for natural history to be purely mechanistic. For example, James Hutton and Charles Lyell, the fathers of modern geology, called for natural history to be caused only by natural laws. Lyell, a mentor of Darwin, became known for his advocacy of uniformitarianism, the idea that the world was formed by a gradual process driven by natural laws. God didn't form the world directly; it arose from secondary causes, his natural laws in action. "The philosopher," proclaimed Lyell, "at last becomes convinced of the undeviating uniformity of secondary causes."[5]

Not only was Darwin aware of these sentiments, but in many ways they helped pave the way for his new theory. To go against uniformitarianism and admit unexplainable discontinuities into his theory would confuse things terribly. Darwin was providing a strictly naturalistic account in accordance with this growing sentiment, and discontinuities would open his theory up to the miraculous.

Today, almost a century and a half later, the situation has changed surprisingly little. Paleontologists have found many new species, giving evolutionists new entries to point to in their evolutionary tree. But strange and seemingly out-of-sequence species have also been found. If Darwin's gradualism were correct, one evolutionist who studied trilobites admitted, "then I should have found evidence of this smooth progression." Instead, he found "unique and advanced" species in the earliest fossil beds. He also found later species simply replacing earlier ones, with no signs of gradual change.[6]

The Fossil Challenge

Consider what today's leading paleontologists say about the fossil record. T. S. Kemp summarizes the situation as follows: "The observed fossil pattern is invariably not compatible with a gradualistic evolutionary process." There is either a problem with the fossil record or with the idea that evolution is gradual. To make the data compatible with the theory, "undiscovered fossil forms can be proposed," or "unknown mechanisms of evolution can be proposed." But neither of these ad hoc hypotheses are known to be true or untrue.[7]

Likewise, Robert Carroll explains how poorly the fossils support Darwin's theory. Our knowledge of the fossil record has progressively increased over the past century, and it "emphasizes how wrong Darwin was in extrapolating the pattern of long-term evolution from that observed within populations and species."[8] The observed pattern of change over geological time scales is nothing like the change postulated by Darwin.

Carroll identifies five specific problems for evolutionists. For example, how did major new structures evolve? "Paleontologists," explains Carroll, "in particular have found it difficult to accept that the slow, continuous, and progressive changes postulated by Darwin can adequately explain the major reorganizations that have occurred between dominant groups of plants and animals."[9] For it is not clear how the minor changes that are observed in species, such as coloration change in moths, could account for the formation of new forms. "How can we explain," Carroll asks, "the gradual evolution of entirely new structures, like the wings of bats, birds, and butterflies, when the function of a partially evolved wing is almost impossible to conceive?"[10]

Another problem is that though we observe a multitude of species, they almost invariably are clustered together. The species "do not form a continuous spectrum of barely distinguishable intermediates. Instead, nearly all species can be recognized as belonging to a relatively limited number of clearly distinct major groups, with very few illustrating intermediate structures or ways of life."[11] Likewise, the fossils fall into a relatively small number of major groups, rather than forming a spectrum of continuous intermediates. "How do we account," asks Carroll, "for the extremely irregular distribution of basic body plans in space and time under a theory of evolution based on gradual and continuous change?"[12]

Another problem is the apparently much greater rates of evolution required during the origin of new groups compared with the lack of evolution thereafter. "If natural selection is continuously acting on all organisms within modern populations," asks Carroll, "why have some animals and plants, such as cockroaches, horsetails, and the horseshoe crab *Limulus*, remained almost unchanged for the past 325 million years, whereas in the same period of time, small, scaly, cold-blooded animals resembling primitive living lizards gave rise to the greatly altered dinosaurs, birds, and mammals?"[13]

Punctuated Equilibrium

Evolutionists have answers and explanations for all of these quandaries, but they are not compelling. The fossil evidence does not naturally lead us to evolution; rather, evolution provides an interpretive filter for the data.

Though the fossil record now supports evolution no better than it did in Darwin's day, one profound change that has occurred is that evolution has attained utter dominance. Whereas Darwin needed to advance his theory, today's evolutionists merely need to rationalize the theory. Darwin could not admit discontinuities into his explanation for fear of failing to advance evolution. But more than a century later, with evolution firmly in place, it was safe for evolutionists to finally acknowledge the fossil record's discontinuities.

In the spirit of Huxley, punctuated equilibrium acknowledges that the fossil record is one of quick change followed by stasis. Proposed in 1972, it codifies ideas that had been around long before and makes it now acceptable for evolutionists to speak of jumps in the evolutionary process.[14] Perhaps those jumps are nothing more than Darwin's gradualism in high gear. Or perhaps the jumps require a different mechanism. In any case, the process is so fast that it leaves no fossil evidence in its wake.

Whereas Darwin criticized the fossil record as being incomplete, evolutionists now have a theory that explains the fossil record. Yes, the record is incomplete, and now we can explain why. But the explanation is still vague, lacking the precise details of just how this rapid speciation would take place and importantly, how anything beyond minor evolutionary change could occur. Ideally, it would be nice to have experimental evidence of the process; however, experiments duplicating the actual process are impossible, since it is supposed to occur over a geologic time scale.

Today's explanations reconcile evolution with the fossil record, but they also reveal the weakness of the evidence. The evidence doesn't stand on its own, but rather needs a clever explanation. For the evolutionist this is an example of scientific progress—the refinement of a theory to better explain the data. But to the skeptic these explanations are patches—just-so stories that cover over an obvious problem. The fossil record doesn't fit evolution, so evolution, we are told, must exert bursts of creative energy. The process is so fast that it conveniently leaves no fossil evidence.

Given these drawbacks, one might think evolutionists would go easy in claiming the fossils as strong evidence for evolution. If these count as strong evidence, then evolution must be weak on supporting evidence. But many evolutionists do lean heavily on the fossil record.

Despite the clear message that Darwin's theory fails to explain the fossil record, evolutionists arrange certain fossils in hypothetical sequences and claim them as overwhelming evidence. The fossil record may be spotty, evolutionists admit, but they argue that in some cases it does reveal evolutionary sequences. An early favorite was the horse sequence, which evolutionists depicted as a neatly arranged lineage of four-legged creatures growing in size and complexity, leading to the modern horse. More recent examples have been the reptile-to-mammal and whale sequences.

Each of these examples has been presented with great confidence in an untold number of textbooks and museum exhibits. But the diagrams are overly simplistic. The meticulous evolutionary lineages that are illustrated mask a confusion of complicating details. Too often the fossils are presented unrealistically as strong evidence for evolution. Evolutionist Henry Gee admits:

> Many of the assumptions we make about evolution, especially concerning the history of life as understood from the fossil record, are, however, baseless. The reason for this lies in the scale of geological time that scientists deal with, which is so vast that it defies narrative. Fossils, such as the fossils of creatures we hail as our ancestors, constitute primary evidence for the history of life, but each fossil is an infinitesimal dot, lost in a fathomless sea of time, whose relationship with other fossils and organisms living in the present day is obscure. Any story we tell against the compass of geological time which links these fossils in sequences of cause and effect—or ancestry and descent—is, therefore, only ours to make. We *invent* these stories, after the fact, to justify the history of life according to our own prejudices.[15]

There is a variety of species in the fossil record, but they persist unchanged. Did the earlier species evolve and give rise to the later species? Perhaps, but the fossil record does not point to such an extraordinary claim. In fact, there remain large gaps even in these carefully selected examples, and explanatory devices such as convergent evolution (different species converging to produce similar species) must be used in liberal doses.

Comparative Anatomy

If evolution is true, then there should be similarities between related species. This is because, with evolution, a new species is not a completely new design, but a derivative of its ancestors. Sometimes a similarity runs through a wide range of species. A favorite example of evolutionists is the pentadactyl pattern—a set of five bones found at the end of the limb structure of frogs, birds, humans, whales and bats, to name a few. The pattern varies widely, but evolutionists argue it is a sign of evolution.

41

Just as the wake of a ship reveals where it has traveled, Darwinists believe the pentadactyl pattern reveals the path that evolution took. The only criterion for passing through the evolutionary process is reproduction. Aesthetics, or the lack thereof, have no bearing. The repetitive pentadactyl pattern is not elegant or optimal, they argue, but it provides the sort of framework that might persist through a variety of species. Patterns such as these, however they arose, are called homologies.

In addition to the visible homologies that have always been obvious to naturalists, we now know of many biochemical homologies. A favorite example is the universal genetic code, also called the DNA code. Although slight variations of the code have been found, the basic code is found in all species—bacteria, plants, and animals. Clearly, there is a great similarity running through the species. For Darwinists this is the ultimate homology amongst the species.

Darwin said he would adopt evolution even if comparative anatomy were its only supporting evidence. Today's evolutionists say the comparative anatomy evidence is an important pillar upholding the fact of evolution. Given these lofty endorsements, we must have a closer look at this evidence. What is it about comparative anatomy that would persuade Darwin, and today's evolutionists as well, that such an unlikely idea was in fact true? Why would similarities between species lead Darwin to propose the remarkable claim that they somehow evolved from one another?

There are many examples of these sorts of similarities, but why does this mountain of evidence so strongly support evolution? What Darwin and his disciples overlook is that this evidence can be used against evolution every bit as much as it can be used to support evolution.

Evolution Explains Too Much

First of all, the evidence must be put in perspective. Remember, Darwinists say that evolution created all life. Consider the tremendous power that we must invest in evolution. It created all the species inhabiting our planet's various environments. In the deep sea the water pressure is tremendous and there is no light. Yet there is life in this hostile environment. Closer to the surface we find a tremendous diversity of lifeforms. Jellyfish drift like ghosts, and whales shoot streams of water into the air.

In fresh water there are fish that use electrical fields to sense and even to stun their prey. Where the water meets the land we find another menagerie of species. Starfish, crabs, anemones, barnacles, and mussels are a few of the creatures in this zone of alternating wet and dry.

On land there are snakes with forked tongues and heat sensors. The forked tongue isn't just a sign of deception. Each fork smells the snake's prey, and when both forks report the same strength of smell the prey is directly ahead. Likewise, the heat sensors tell the snake the direction of prey. The snake wags its head back and forth to map out what's ahead.

In colder environments insects undergo a preflight warm-up. Their muscle temperature is increased by shivering before taking flight. From the mountains to the desert to the frozen tundra, life abounds in seemingly endless variety and contrivance.

Evolution, if true, has truly incredible creative powers. This is because, Darwinists explain, the evolutionary process selects the right designs for the right environments. But when similarities are found, Darwinists chalk them up to evolution's settling for anything that happens to work. Suddenly evolution doesn't work so well.

It seems that evolutionists are having it both ways. Highly efficient designs that leave engineers in awe are due to evolution's selective power, and any repeated patterns are the leftovers from the process. There doesn't seem to be anything that evolution cannot explain. Every time a new design is discovered, it is said to be a result of evolution, whether we find it to be optimal or gawky.

Vestigial Organs

Those gawky designs are a popular source of evidence for evolutionists. The designs make no sense, evolutionists say, so they must have been formed by evolution. For example, evolutionists believe that as species evolve, certain organs or structures may no longer be needed. It may take time before the organ is completely phased out by the evolution. In the meantime it remains as a vestige of the evolutionary process.

But how can we be so sure that such vestigial organs really lack function? Evolutionists believe life was created by blind, unguided forces. For this reason they are prone to interpreting nature as haphazard. If a particular design is not understood or is confusing, they will more often see it as a result of evolution's unguided process than as a gap in our knowledge. If no function can be found, then an organ or a structure is likely to be labeled as vestigial.

One Cannot Prove an Organ Is Vestigial

It is difficult to measure the fitness of supposedly vestigial organs, and many designs that evolutionists once considered to be useless have been found to have a clear function. The problem is that in order to identify an organ as vestigial, we need to measure its adaptive value—its contribution to the production of offspring.

At the core of evolutionary theory is the idea that in most instances, it is the fittest that reproduce. But due to the complexities of nature and its life-forms, we usually cannot measure fitness, aside from counting the offspring. Those organisms that leave more offspring are usually more fit, but we are not sure precisely why.

Thus, it is difficult to show that a particular organ lacks value. Whether we are talking about an organ that is thought to contribute little to the overall fitness or is thought to be inefficient, our failure to find positive value does not imply that it is nonexistent. One cannot conclude something does not exist unless one has looked in all possible places at all possible times. In fact, the claim that an organ is vestigial can only be rejected. When we find that the organ makes a positive contribution to fitness, then we disprove the vestigial claim, but it is practically impossible to prove the claim by failing to find such a contribution. It is not surprising, therefore, that the history of vestigial organs is one of shrinking lists.

In 1895 Robert Wiedersheim published a list of eighty-six organs in the human body that he supposed to be vestigial. The vast majority of the organs he claimed to be vestigial are now known to be functioning organs. For example, there was the pineal gland, which is now known to be part of the endocrine system that sends chemical messages (hormones) in the blood and interacts with the nervous system. Wiedersheim also claimed the coccyx, a short collection of vertebrae at the end of the spine, was vestigial. But the coccyx is the attachment point for several important muscles and ligaments. And Wiedersheim claimed that the thyroid and thymus glands and the appendix were vestigial, but important functions for all three have since been discovered.

The thyroid gland consists of two lobes on either side of the windpipe and produces thyroxine, which regulates cellular metabolism. It is important in cold temperatures and in child growth. The thyroid gland also produces calcitonin, which helps regulate blood calcium levels. Its malfunction and enlargement—the condition known as goiter—is visible as a swelling of the front of the neck. Both the thymus gland and the appendix contribute to the body's immune system. In 1981, zoologist S. R. Scadding analyzed Wiedersheim's claims and had difficulty finding a single listed organ that was not functional, although some only in a minor way.[16] He concluded that the so-called vestigial organs provide no evidence for evolutionary theory.

In recent years the term "junk DNA" has been used to describe portions of the genome that evolutionists believe are useless relics of evolution. These segments of DNA were considered to be useless because the same sequence of nucleotides is repeated over and over. But again subsequent research has found that there indeed appears to be a reason for these "vestiges." They serve as binding sites for bridging proteins that assist in the DNA replication process. After the DNA replicates, the cell divides. Each daughter cell contains the same copy of the original DNA.

There is much "junk" DNA in the genome and, as one researcher put it, "this makes sense if one of their roles is to bind to the bridging proteins . . . to keep the replicated DNA sisters together until it is time for them to separate. Multiple bridging sites throughout the DNA would be needed for this system to work. They couldn't be unique sequences." In other words, there is a reason why the same sequence is repeated over and over. It is not evidence of evolution, it is evidence of design.[17]

Evolution Is the Evidence for Vestigial Organs

If the fitness value of an organ is difficult to measure, then how does it qualify as "vestigial"? Even the simplest of organs have many processes and structures that are still not understood, and cannot be monitored continually. Nor is it possible to construct and evaluate alternative versions that might be better adapted. No one has even come forth with an alternate design for an organ that is thought to be vestigial.

In fact, evolutionists do not require that an organ lack function in order for it to be vestigial. For example, evolutionist Sir Gavin De Beer cites insect wings that must have evolved into gyroscopic navigational devices, and muscles that must have evolved into electric organs for signaling or attacking. This calls for a certain amount of serendipity, as the blind, unguided process of natural selection is supposed to remake an outmoded structure into something highly complex and quite useful.

The upshot is that evolution accommodates all possibilities. Whether a function is discovered or not, the structure can still be safely labeled as vestigial, and as De Beer concludes, even the useful ones "are evidence for evolution."[18] Likewise, evolutionists Edward Dodson and Peter Dodson explain:

> When structures undergo a reduction in size together with a loss of their typical function, that is, when they become vestigial, they are commonly considered to be degenerate and functionless. [George] Simpson has pointed out that this need not be true at all: the loss of the original function may be accompanied by specialization for a new function. Thus the wing of penguins has become reduced to a point that will not permit flight, but at the same time it has become a highly efficient paddle for swimming. The wings of rheas, ostriches, and other running birds are also much

reduced, and have been described as "at the most still used for display of the decorative wing feathers." But Simpson has observed that the rheas, when running, spread the wings and use them as balancers, especially when turning rapidly. It seems quite probable that this is true of the running birds generally.[19]

What, then, defines a vestigial structure? If a penguin's wing is highly efficient for swimming, then why should we think it is vestigial, aside from simply presupposing it was formed by evolution? The idea that vestigial structures can, in fact, be perfectly useful makes the argument subjective. A character trait that is fully functional for one observer may be only partially functional for another observer, and may be considered inefficient for yet another observer.

Tim Berra tells us that the small bones found in the rear quarters of whales and snakes are "surely of no value" and that this supports "the evolutionary explanation that whales evolved from terrestrial mammals, and snakes from lizards."[20] And Mark Ridley assures us that the recurrent laryngeal nerve "is surely inefficient."[21] But there is no scientific evidence to back up these claims.

The very use of the term *vestigial* begs the question, for vestigial structures serve as evidence for evolution only if they are indeed vestigial. But we cannot know they are vestigial without first presupposing evolution, because we cannot directly measure their contribution to the organism's fitness. Therefore, when evolutionists identify a structure as vestigial, it seems that it is the theory of evolution that is justifying the claim, rather than the claim justifying the theory of evolution. As J. B. Meyer cogently observed ten years after Darwin first published his theory, Darwinism is not so much a hypothesis proposed to explain facts as an invention of facts for the support of a hypothesis.[22]

Amazing Patterns That Cannot Be Homologies

Another problem with the comparative-anatomy evidence is that nature is full of similarities that do not fit the evolution pattern. For example, there are two species of frog whose eyes are similar but are formed differently. Did evolution preserve the end product but not the development path? This seems unlikely, but evolution's other alternative—that the eye evolved independently in these two cousin species—is even more unlikely.

Another well-known example is the marsupial-placental convergence in mammals. In marsupials, the young are born soon after conception and continue their early development in a pouch on the mother's belly for weeks or months. In the placentals the embryo is attached to the uterine wall and grows considerably before birth.

All mammals are believed to have evolved from a small, four-footed animal something like today's mouse or shrew. The placentals and marsupials are supposed to have formed an early division in the ancestral lineages derived from this humble creature. Somehow this creature duplicated itself into two different versions—one marsupial and one placental. After that, the marsupials found homes in South America, Australia, and even North America, while the early placentals found homes mostly in the Northern Hemisphere.

The marsupial and placental lineages include great variation. There are hundreds of distinctive types, including the bat, the whale, the wolf, and the rodent. There are creatures that fly, glide, climb, swim, dig, and graze. Some are plant-eaters, some are meat-eaters, and some are both. But amongst this tremendous diversity, there are uncanny similarities.

For example, there are the striking similarities between the marsupial and placental saber-toothed carnivores—*Thylacosmilus* of South America and *Smilodon* of North America. Both sport the same distinctive stabbing upper canines and a protecting flange of bone on the lower jaw.

The marsupial known as the "flying phalanger" and its North American counterpart, the flying squirrel, have distinctive coats that extend from the wrist to the ankle, giving them the appearance of a living hang glider. This, along with their bushy tails, gives them their gliding abilities.

Then there is the marsupial mole of Australia, whose body plan and behavior are like those of the moles of the Northern Hemisphere. Both have enlarged forelimbs and reduced eyes for their subterranean environment. The marsupial and placental wolves have a very similar skull shape and body form. And there are more. One can find marsupial counterparts to the placental rats, anteaters, cats, and mice.

All these "cousin" species have their differences, especially in their young-bearing habits. But the repeated duplication of distinctive characters is remarkable. Nonetheless, evolution must consider these to be cases of convergent evolution. The common ancestor of the marsupial and placental lineages is supposed to have somehow speciated to create the prototypical marsupial and placental forms. Then these two species went their separate ways. Separated by continents, they gave rise to remarkably similar lineages.

The similarities between the marsupial and placental cousin species are certainly as significant as those homologies Darwinists claim as evidence for evolution. The difference is that the marsupials and placentals cannot be part of the same evolutionary lineage. Evolutionists are forced to explain them as a remarkable example of convergent evolution.

There are also amazing analogies at the molecular level. The textbook example of this is subtilisin, another serine protease like those we examined in chapter 3. As with the other serine proteases, the function of subtilisin is intricate. It is assembled so that three of its amino acids, which are initially far apart, come together in a precise geometry to enable fast and efficient enzymatic reactions. The reactions are characterized by several key functions that act in concert. Subtilisin falls into the serine protease category because of its function, but its amino acid sequence and structure are significantly different from the other serine proteases. Subtilisin is sufficiently different from the others that it is generally thought to have evolved independently.[23] Another amazing case of convergent evolution.

Of course, evolution can explain all of this, but the explanations go no further than sweeping generalities. As Tim Berra concludes, "such close similarities in very unrelated groups are easily explained as a result of convergent evolution."[24] Perhaps too easily. Though evolution is a blind process that produces a broad menagerie of species and designs, it also is supposed to produce striking similarities.

This duality causes a tension within Darwinism. On the one hand, the evolutionary process is supposed to be unguided. Many Darwinists have argued that evolution is not very repeatable. If history were to be repeated again under slightly different conditions, they claim, then the evolutionary process would produce a different set of species. This view emphasizes a fundamental principle of Darwinism: that the evolutionary process is random and unguided. But this view has difficulty accounting for the striking similarities found among the species.

Therefore, other Darwinists argue that evolution naturally converges to certain designs. These preferred designs, they say, are the result of natural laws and the workings of biology. This view more easily accounts for nature's striking similarities—it clearly seems to be more consistent with the evidence at hand. But this view is less consistent with Darwinism's fundamental principle that the process is unguided. Can we really believe that Darwin's unguided process would produce such striking similarities in different lineages?

As we shall see in later chapters, the idea that evolution is unguided is crucial to Darwinism. Hence, while the evidence may suggest that there are preferred designs in nature, it must not be allowed to suggest that evolution is anything but the blind interplay of natural forces.

This tension is not particularly troublesome for Darwinists, because evolution can explain just about anything. Very different species, very similar species, and anything in between can be explained one way or another, and this is the point. The comparative anatomy evidence does not lead us to evolution—it does not explain evolution. Rather, evolution explains the comparative anatomy evidence—it is an interpretive filter telling the Darwinist whether a similarity reveals a common ancestor or convergent evolution.

5

Straining at the Gnat

The Scientific Evidence for Evolution, Part 2

The traditional evidences for evolution are not very convincing. In fact, under scrutiny they often reveal additional problems with the theory. In this chapter we continue analyzing more of the claims of evidence for evolution.

Molecular Comparisons

Molecular Clock

In the 1960s molecular biologists learned how to analyze protein molecules and determine the sequence of amino acids that make up a protein. It was then discovered that a given protein molecule varies somewhat from species to species. For example, hemoglobin, a blood protein, has similar function, overall size, and structure in different species. But its amino acid sequence, while similar, is not identical. Emile Zuckerkandl and Linus Pauling hypothesized that such sequence differences were the result of a relatively constant rate of evolutionary change occurring over the history of life and could be used to estimate past speciation events— a notion that became known as the *molecular clock*.[1]

For example, if two species have hemoglobin proteins that are almost identical, then evolutionists infer that the two species have a recent common ancestor on the evolutionary tree. Only recently did the two species diverge, because their hemoglobin proteins are so similar. On the other hand, if the two hemoglobin proteins have many differences, then evolutionists believe that the two species have been evolving independently for a longer time. The most recent common ancestor of the two species would be lower on the evolutionary tree.

The molecular clock has been extolled as strong evidence for evolution. The National Academy of Sciences claims that the molecular clock "determines evolutionary relationships among organisms, and it indicates the time in the past when species started to diverge from one another."[2] Or as a leading molecular evolutionist wrote, the comparisons of the hemoglobin proteins in different vertebrates are *"only* comprehensible within an evolutionary framework."[3]

But how good is the molecular clock? The evolution literature is full of instances where the molecular clock is apparently inaccurate. For example, it was found early on that different types of proteins must evolve at very different rates if there is indeed a molecular clock. Furthermore, it was found that the evolutionary rate of certain proteins must vary significantly over time and in different species.[4] The snake cytochrome c protein, for example, must have evolved several times faster than in other species.[5] It was also found that the molecular clock doesn't seem to work very well for bacteria. The molecular data make closely related bacteria look like distant relatives—as different as insects are from mammals, for example.[6] It was also found that the relaxin protein was anomalous when compared across different species. The pig, for example, was found to be more closely related to a shark than to a rodent.[7] "The conclusion to be drawn from the relaxin sequence data," wrote one researcher, "is that they do not fit the evolutionary clock model."[8] Furthermore, in order to fit the data to the molecular clock hypothesis, one must imagine that different regions of the genome evolve at different rates for a species, and that the same region evolves at different rates in different species.[9] Many instances have been discovered where the data do not seem to fit the molecular clock model, and some researchers wonder if this is where the real message is:

It seems disconcerting that many exceptions exist to the orderly progression of species as determined by molecular homologies; so many in fact that I think the exception, the quirks, may carry the more important message.[10]

There are, to be sure, explanations for the various anomalies. In fact, evolutionists have many possible reasons why data might not fit the molecular clock. For example, the molecular clock model depends on the species population size and likely on its life span—the generation time effect. Also, varying DNA replication accuracies in different species may affect the clock; or could it be possible that the protein under study has somehow changed its function in its evolutionary history? On the other hand, perhaps a horizontal gene transfer has taken place at some point in the organism's history; or perhaps so many mutations have occurred that the picture has become blurred. Perhaps molecular evolution experiences elevated rates during periods of adaptive radiation or maybe slightly deleterious mutants were incorporated during population bottlenecks. Early skeptics criticized evolution as a theory that explains everything yet is in need of so much explaining.[11] More than a century later little has changed.

It might seem that, given this battery of explanatory devices, there would be no observation that evolutionists could not explain. But there are, in fact, some cases that remain difficult to explain. There is, for example, the serum albumin gene family, which shows significant deviations from clock-like evolution. Researchers who investigated these genes concluded that the molecular clock "is subject to the same vagaries as the rest of biology. Models are only models; they are only as good as the underlying assumptions. And if the number of assumptions (unknowns) is greater than the number of equations, a rigorous solution is but an illusion. This seems to be the case with the molecular clock."[12] Other erratic molecular clocks include those based on the superoxide dismutase (SOD) and the glycerol-3-phosphate dehydrogenase (GPDH) proteins. On the one hand, SOD unexpectedly shows much greater variation between similar types of fruit flies than it does between very different organisms such as animals and plants. On the other hand, GPDH shows a more or less reverse trend for the same species. As one scientist concluded, GPDH and SOD taken together leave us "with no predictive power and no clock proper."[13]

Another problem with the molecular clock hypothesis is that the clock must be calibrated before it can be used. Evolutionists use fossil data to calibrate the clock. Fossils are used to reconstruct a hypothetical evolutionary tree—a phylogeny—including the geological time since particular speciation events. The molecular differences between species are calibrated to those speciation events. The molecular clock is then used to measure the time since other speciation events. In many cases the molecular clock conflicts with the fossil data. That is, once the clock is calibrated using one part of the fossil record, the clock then conflicts with another part of the fossil record. The molecular and fossil data often do not give similar results. This can be explained, for example, by pointing to possible errors in the fossil record. But the problem in all this is that the molecular clock is being calibrated with the same sort of data that the clock then contradicts. As one researcher put it: "Even if one makes the bold assumption that molecular clock models have little error, there seems little objective reason for accepting a few fossil dates used in calibrations and rejecting as unreliable the much more numerous fossil dates that contradict the resultant molecular estimates."[14]

Yet another problem with using the fossil record for calibration is that the speciation events are themselves not in the fossil record. That is, the purported speciation events are inferred from the fossil record assuming evolution is true. They cannot be measured directly because the fossil record lacks the temporal resolution required to measure such events if they had actually taken place. The fossils tell us something about what the ancient life-forms looked like; the fossils do not tell us how those life-forms came about. The molecular clock hypothesis ultimately relies on the presupposition of evolution—it is not a new and independent source of evidence *for* evolution.

There are plenty of anomalies that challenge the molecular clock hypothesis, and it has been controversial practically since it was proposed. While evolutionists continue to employ the hypothesis, its validity remains in question. No doubt, for those who believe in evolution the molecular comparisons are another piece of information that must be considered. But it is a considerable stretch to claim the molecular clock as strong evidence for evolution.

The Hierarchical Pattern

Since antiquity it has been known that nature's species can be grouped. The species are not randomly or arbitrarily designed. Nor are the species uniformly spaced out. Nor do they form simple clusters. Instead they form clusters upon clusters. As Darwin described it, groups of species are "clustered round points, and these round other points, and so on in almost endless cycles."[15]

In the eighteenth century the Swedish botanist Carl von Linné, or Linnaeus, sought to organize and classify all the world's species. Linnaeus eventually used seven hierarchical levels in his system to describe the master plan of the wise Creator.

At the highest level all the world's species were divided into two kingdoms: plants and animals. These were then divided into several phyla, which in turn were divided into several classes, and then orders, families, genera, and finally species.

The Linnaean system captures groupings that really exist in nature's species. Everyone knows that species are distinct from one another. But the species indeed tend to cluster into genera. And the genera tend to cluster into families, which tend to cluster into orders, and so forth.

Evolutionists claim this distinctive pattern as strong evidence for their theory. In fact, some say it is a necessary outcome of evolution.[16] Of course the Linnaean pattern does not *prove* evolution is true, but if the pattern was not found in nature, they say, then evolution would be falsified.

The Linnaean pattern is considered to be strong evidence because it seems to naturally result from the evolutionary process. If you have a gradual process of evolution, then cousin species will be more similar to each other than to the species that are on the next branch over on the evolutionary tree. But species on those two branches will form a group—they will be more similar to each other than to species on another major limb of the tree. The overall result is the familiar nested clusters—the Linnaean hierarchy. Though the existence of the hierarchy was obviously known long before Darwin formulated his theory, it is nonetheless claimed as strong evidence because Darwin's process predicts it to occur.

But does it really? Is the Linnaean hierarchy really the only possible result if evolution is true? The most obvious problem with this claim is that it works only for gradual evolution. If nongradual mechanisms occur (and evolutionists have contemplated them to explain the fossil record), then the hierarchy need not occur. It also need not occur if the evolutionary rate is high. We saw above that high rates of evolution are used to explain anomalies in the molecular clock data. High rates are also used to explain proteins that do not fit the Linnaean hierarchy. Those proteins evolved so fast, say evolutionists, that they left no trace of the hierarchy.

High rates are also used for proteins at the other end of the spectrum that show little variation. These proteins do fit the Linnaean hierarchy, though for the most part they are identical from species to species. If evolution is true, then these proteins must have changed very slowly over time. For this reason evolutionists use them to look far back into time, for according to evolution, they retain information from long ago.

Why do these proteins evolve so slowly, while others evolve at a high rate? The reason for this, evolutionists say, is that these proteins are highly constrained. Natural selection will pick out only a few variations of these proteins because the other variations don't work well enough. But if this is true, then how did these proteins arise in the first place? If they show no flexibility in their design, then how could they have gradually evolved? It seems that they must have appeared in their present form, but the odds on this occurring are miniscule. Evolution's approach is a gradual buildup, not an instantaneous creation.

The answer to this problem, say evolutionists, is that these proteins experienced a high rate of evolution long ago, because they were less constrained. Once again, high rates of evolution are used to explain the data. And if a high rate of evolutionary change is used to explain so many observations, it could also be used to explain the absence of a nested hierarchy.

In other words, evolution has been formulated to explain the Linnaean hierarchy, but it could also be formulated differently. The point here is not that evolution fails to predict or explain the present situation. The Linnaean hierarchy and its variations can be explained by evolution perfectly well. The point is simply that evolution can explain a great many outcomes. It is not unusual for theories to have this sort of flexibility, at least in some regards. But when a theory can explain multiple outcomes, it cannot claim any one of the outcomes is strong evidence for the theory.

The Same Phylogeny (Evolutionary Tree) from Different Design Features

Early in the twentieth century scientists studied blood immunity and how the immune reaction could be used to compare species. Not surprisingly, the blood studies produced results that parallel the more obvious indicators, such as body plan. That is, in terms of the immunological distance between species, humans are more closely related to apes than to fish or rabbits, just as they are in terms of the visible features. The evolutionary tree, or phylogeny, derived from the immunological distance data is similar to the traditional phylogeny. Ever since this discovery evolutionists have claimed it is convincing evidence for evolution.[17]

Today, it is more common for evolutionists to refer to protein sequences than to blood immunity, but the reasoning is the same. The idea is that the phylogeny we find by arranging the species according to the visible traits is the same as the one we find when arranging by molecular traits. Species that have similar bones or body plans will also tend to have similar proteins, and species that differ in those visible features will tend to have greater differences in their proteins.

55

These results are not surprising, for the visible features are ultimately driven by the molecules. The information stored in the DNA determines both the proteins and the body plan. In other words, we would expect the visible features to be correlated with the molecular features.

But what if this is not the case? What if molecular features need not correlate with the visible features? Perhaps similar species could arise from different molecules and vice versa. Our understanding of biology is limited, so we don't know for sure, but there is evidence for this. For example, some proteins seem to work just fine when transplanted from one species to another. What if this is generally true? What if, for example, a frog would function just fine even if its hemoglobin and other proteins were from practically any species? Then we would have no reason to think that molecular features should correlate with visible features; and importantly, we would have no reason to expect the phylogenies from different features (such as hemoglobin and the visible features) to be similar. The visible features form the Linnaean hierarchy, but the hemoglobins might be random.

Evolutionists claim that the similar phylogenies that come from different features constitute strong support for their theory.[18] Implicit in their claim is the assumption that molecular features need not correlate with visible features. The fact that they do correlate is seen as a great coincidence for which evolution is the obvious explanation. This line of thinking may be reasonable, but it hardly makes for good evidence. There is a host of reasons why molecular features might need to correlate with visible features; we simply do not know enough about biology to say for sure.

Another reason why this evidence is weak is that the correlation between molecular and visible features is often not very good. If a high correlation makes for powerful evidence, then what about the many exceptions that fill the literature? Often the molecular data, when compared across many species, do not form a phylogeny like that of the visible features.

For example, molecular studies of bats have challenged the traditional phylogeny. Some species of bat have an incredibly complicated echolocation system that tracks objects as small as a mosquito by sensing the echoes of the bat's own squeaks. The bat emits a high-pitched squeak, well beyond the range of human hearing, up to 2,000 times per second. It determines both range and direction to the tiny mosquito by sensing the echo while filtering out echoes from the squeaks of nearby bats.

Beyond making general speculation, evolutionists cannot describe how such a system would have evolved, but it is usually assumed that such a unique and complex system evolved only once. The different species of bat that have echolocation are all thought to derive from the ancestor that first developed the system. But now the new molecular phylogenies call for a different arrangement. If they are correct, then echolocation must have evolved more than once, independently, in different bat species.[19]

Another striking example of mismatch between molecular and traditional phylogenies comes from mitochondrial DNA. Most of the cell's DNA is in the cell nucleus, but the mitochondria also have some DNA that codes for a relatively small number of proteins. A recent study found that the mitochondrial DNA provided a statistically high-confidence phylogeny that "was clearly the wrong answer." For example, frogs and chickens were clustered with fish.[20]

In addition to conflicting with phylogenies based on visible features, molecular phylogenies are also sometimes internally inconsistent. For instance, a study of 188 different genes from five different light-harvesting bacteria, each from a different phyla, showed dramatic inconsistencies. Instead of the 188 genes pointing to a particular phylogeny they showed no strong preference. In fact, every conceivable phylogeny found support amongst the 188 genes. One could argue for completely different evolutionary histories depending on which genes one selected— the different design features did not converge to the same phylogeny.[21]

The point is not that evolution cannot explain these anomalies. In fact, evolutionists have a variety of explanatory mechanisms. For instance, the molecular data may have diverged sufficiently that the comparisons between them are not valid. Or there may be too much noise in the molecular data. Perhaps a massive dose of lateral gene transfer has altered the molecular data.

On the other hand, the molecular data may be biased by the particular set of species under study. It turns out that molecular phylogenies are sensitive to the choice of species in the data set. One solution for this problem is to increase the number of species analyzed; however, in addition to dramatically increasing the computation time required, a larger data set can increase the chances of obtaining an erroneous phylogeny.[22]

Yet another explanatory mechanism for the molecular-visible feature mismatch is that the molecular data ought not to be equally weighted. Perhaps those segments of the molecular data that influence protein function should be assumed to be more important than the other segments. Or perhaps segments influencing the protein structure should be more heavily weighted. One study found that those segments associated with the oily amino acids should be deweighted.

Evolution has any number of explanatory devices with which to reconcile differences between the molecular and visible features and otherwise refine the molecular results as needed. The point is not that evolution cannot explain the mismatches we observe, but rather that with its various mechanisms evolution can explain a variety of results, including molecular phylogenies that match the traditional phylogenies and those that do not.

Finally, while evolutionists are quick to point out similarities they think share a common evolutionary history, they rarely point out the other similarities that do not fit the evolutionary picture. There are many striking similarities in biology that appear in otherwise distantly related species.

We saw examples of these "amazing patterns" at the end of chapter 4. We will see another example, from the mouse and human genomes, in the next section. If evolution is true, then these similarities must have arisen independently, but they are strikingly similar. Furthermore, biology is full of designs that are not similar—even in cousin species.

But none of these counter cases poses a problem for evolution. So while evolutionists claim their theory predicts similar designs in similar species, it also can predict similar designs in very different species, and different designs in similar species.

Genomic Similarities

In recent years the entire complement of genes (the genome) of several species has been mapped out. Evolutionists are using these genome data to refine their theory. They are also making some high claims. The genome data sets, say evolutionists, are adding striking new confirmations for their theory. Let's have a look.

One evidence evolutionists point to is the high similarity between the human and chimpanzee genomes. The DNA coding regions have been estimated to be 98.7 percent the same, and this shows how easily the human could have evolved from the chimpanzee. In fact, the differences amount to a few genetic changes of the sort that we have observed to take place.

The problem with this reasoning is that these coding region differences are too small to account for the differences between the human and chimpanzee. We have indeed observed these sorts of genetic changes, and they either cause problems or at best result in little or no change. They do not induce anything close to the differences between these two species. Furthermore, we are now finding that the human and mouse genomes are strikingly similar. It appears that mice and people share essentially all the same genes. Clearly, the construction of organisms depends on more than just the DNA coding regions.

What all this tells us is that the mapping from coding region difference to species difference is not straightforward. A relatively small difference in the genome coding sequences can translate into a significant difference between the species. Clearly we do not have a good understanding of how the genome works.

In fact, evolution does not predict the human and chimpanzee genomes to be 98.7 percent similar. If they were, say, 90 percent or 80 percent similar, or even less, then this would not cause a problem for evolution. Indeed, we are now finding that 98.7 percent is an overestimate and that there are more differences between the human and chimpanzee genomes than were previously thought. When evolutionists claim the human-chimpanzee genome similarity is evidence for evolution, they are leaving out a host of complicating details. What these data reveal is how little we understand the development of a species, not that evolution is confirmed.

The next evidence claimed by evolutionists is that evolution predicts the coding regions to be more similar, and the noncoding regions to be less similar. Why? Because mutations are usually harmful, so they are weeded out in the coding regions. The noncoding regions, because they are thought to serve no purpose, do not weed out the mutations and thus evolve at a higher rate.

Of course, this prediction hinges on the noncoding region having no function. Furthermore, this is not really a prediction, because evolution would have no problem if this were not the case. For example, if the genes had greater variation than the noncoding regions, then evolutionists could say selective forces drove the gene differences, while the noncoding regions have some function that limited the amount they could evolve.

In fact, recent comparisons between the mouse and human genomes have uncovered this very situation. That is, some noncoding regions in the mouse and human genomes have been found to be practically identical. Not surprisingly, there isn't so much as a hint that evolution should be questioned. Instead, rate variations and the possibility that some noncoding DNA is "somehow helpful to the genome after all" are cited as explanations. As one researcher put it, "we're seeing there's more to the story" than we realized. "Genomes are evolving," another added, "in a completely nonuniform way."[23]

It is striking how confident evolutionists are in light of such great uncertainties about how the genome works. They transform evidence that is ambiguous or even questionable into high claims for their theory. I have repeatedly seen evolutionists claim that the new genome data provide compelling evidence for macroevolution, that a highly variable noncoding region that appears at the same place in the mouse and human genomes, for example, indicates an evolutionary relationship. But why? Evolution can explain just about any situation. Whether the noncoding region is highly variable, completely identical, anywhere in between, absent from one of the genomes, or absent from both, evolution can come up with an explanation.

Small-Scale Evolution

Evolution is sometimes called a historical science as opposed to an experimental science. The evolution "experiment" was run once and only once—it cannot be repeated in the laboratory. Therefore, evolution relies on historical evidences, such as the fossil record, and on circumstantial evidence, such as comparative anatomy.

But evolution on a small scale is observable. There are, for example, the many examples of our domesticated plants and animals. Breeders have selected for desirable traits and thus "evolved" everything from corn to cattle. Over a century of corn breeding has increased corn oil production dramatically. Evolution also produces unwelcome guests, such as bacteria that gain resistance to antibiotics or insects that have become resistant to pesticides.

We can also find examples of small-scale evolution in nature. A well-documented case is the variations in the beaks of finches caused by drought on the Galapagos Islands. Biologists studied the birds in detail from the 1970s to the 1990s. They found a 4 percent increase in the average beak size of one species after the drought of 1977. The beaks later decreased, after years of heavy rains. No overall trend was found, but small-scale evolutionary change was observed.

Even more dramatic are the changes observed in viruses and bacteria. At this end of the spectrum, these tiny organisms are able to evolve much more rapidly than larger, multicellular organisms. For example, experiments have shown that bacteria populations can modify their metabolism to adapt to strange environments.

But evolution appears to have its limits, especially in multicellular organisms. This is the first problem with this evidence. Yes, small-scale evolution is a fact, but there is no reason to think it is unbounded. In fact, all our data suggest that small-scale evolution *cannot* produce the sort of large-scale change Darwinism requires. The point is not that such an extrapolation has been disproven—it has not. The point is simply that such extrapolation has little basis in fact. The beaks of the finches, for example, returned to normal when the rains returned to the Galapagos Islands. The small-scale evolutionary evidence hinges on its ability to extrapolate to large-scale change. The lack of evidence for such extrapolation reveals a weakness in this evidence.

Evolutionists often respond to this criticism by asking for proof that small-scale change *cannot* extrapolate to large-scale change. Small-scale change does not appear to be able to extrapolate, but it would be difficult to prove this. In science, however, the skeptic is not required to disprove a theory. Instead, the burden of proof is on the one proposing the theory. The proponent needs to show the theory to be likely. The failure to falsify a theory is of little consolation if the theory is unlikely to begin with.

Macroevolution needs to show that large amounts of change are feasible. Bacteria to fishes to amphibia to reptiles to mammals is a tremendous amount of change compared to small-scale evolution. The organism must be viable and fit at every point along these paths of change. What actually have been demonstrated are (1) many examples of change that lead nowhere evolutionarily, either due to reduced fitness or sterility, and (2) some examples of change that represent a tiny fraction of the required change. Imagine using a flat parking lot to reason that the earth is flat. The question of whether viable Darwinian paths exist between the species remains unanswered even after 140 years. It is an unsolved problem, and it raises a legitimate concern about evolution's validity.

Even evolutionists debate amongst themselves whether macroevolution is nothing more than repeated rounds of microevolution. The large-scale patterns of life, writes one evolutionist, reveals "a richness to evolution unexplained by microevolution."[24]

And as we saw in chapter 4, leading paleontologists agree that the fossil record does not reveal a pattern of accumulated small-scale change. Furthermore, it is not clear from genetics how small-scale change is supposed to contribute to large-scale change. As evolutionist Ernst Mayr admits:

> When we look at what happens to the genotype during evolutionary change, particularly relating to such extreme phenomena as highly rapid evolution and complete stasis, we must admit that we do not fully understand them. The reason for this is that evolution is not a matter of changes in single genes; evolution consists of the change of entire genotypes.[25]

Another weakness is that small-scale evolution requires a complicated reproduction system for the required genetic variation. We saw in chapters 2 and 3 some of the reasons why this system is so complicated. How did this system arise? Did it evolve? If so, then evolution produced a system that in turn fuels evolution. In other words, evolution relies on the preexistence of biological variation without understanding from where its generator came. We now know how variation comes about but not how the machine behind it came about.

Without variation, natural selection was powerless to work, yet somehow a source of variation arose. Evolutionary speculation is alive and well about how this might have happened. For example, perhaps a simple source of variation somehow arose first and later evolved and gained complexity. This is, of course, possible, but we have no evidence for it. The point here is not that small-scale change could not have extrapolated to large changes, but simply that it is unreasonable for this to serve as strong evidence for evolution. As with the fossil and comparative anatomy evidence, the small-scale evolution may seem compelling at first, but it fails to hold up under scrutiny.

The Origin of Life

Darwin's quest was to demonstrate that purely natural mechanisms were sufficient to account for life. Though Darwin focused on the derivation of new species from preexisting species, his vision also included the initial creation of life from lifeless chemicals, and his disciples have taken up the challenge of showing how this might have occurred.

The origin of life has always been a difficult area of research for evolutionists because of the limited evidence at hand and the high complexity of even nature's simplest life-forms. Scientists currently think that about three hundred genes are essential for the most simple bacterium to operate under normal conditions. But the design space of even this minimal organism is immense. Obviously, such an organism could not have spontaneously formed from a pool of chemicals. The challenge for evolutionists is to find simpler, subcellular organisms that could have led up to the first living cell. What did these stepping-stone organisms look like, and how were they formed? There is no evidence even for their existence, so the answers to these questions are speculative.

Two big requirements for these hypothetical subcellular organisms are that there must be a natural mechanism for their creation and that they must not only reproduce but also lead to more complex organisms. To date, evolutionists have failed to construct such a pathway or even a single reproducing subcellular organism. In fact, they have not even described on paper the complete details of any hypothetical pathway. What the research has produced so far is a wide range of competing concepts of how the process might have worked. Do these results constitute evidence for evolution? The National Academy of Sciences (NAS) thinks so: "For those who are studying the origin of life, the question is no longer whether life could have originated by chemical processes involving nonbiological components. The question instead has become which of many pathways might have been followed to produce the first cells."[26]

Such views are not limited to the NAS. Science writer Carl Zimmer states that scientists "have found compelling evidence that life could have evolved into a DNA-based microbe in a series of steps."[27] This is overly optimistic to the point of misrepresenting the state of the research. Origin-of-life research is nowhere near such an achievement. In fact Zimmer's description of the research is liberally sprinkled with qualifiers, such as "might have," "may have," and "scientists suspect."

Origin-of-life research is highly speculative and lacks strong evidence. For many, it seems clear that the research is motivated by the assumption that evolution is true and that the results must be interpreted as such. In other words, researchers will never conclude against a natural origin of life, and results will always be given a pro-evolution spin.

Zimmer writes that the raw materials required for the origin of life could have come from space. "Meteorites, comets, and interplanetary dust," he explains, "could have seeded the planet with components for crucial parts of the cell."[28] But only a few of the many chemicals used in the cell would have been available and only in low concentrations. For this problem, Zimmer explains that the chemicals "might have been concentrated in raindrops or the spray of ocean waves."[29]

If we doubt this particular scenario, Zimmer explains that other scientists "suspect that life began at the midocean ridges."[30] And yet another take on the problem is the possibility of cycles of chemical reactions that scientists suspect could sustain themselves. "There may have been many separate chemical cycles at work on the early Earth. . . . The most efficient cycle would have outstripped the less efficient ones. Before biological evolution," Zimmer easily concludes, "there was chemical evolution."[31]

What, for many scientists, is a very difficult problem far from solution somehow becomes for evolutionists a success story. The success comes not from a compelling solution but from an abundance of speculations, none of which has really solved the problem. Unfortunately, this unwarranted claim of success is characteristic of how evolutionists evaluate the state of evolutionary research.

It is a fundamental tenet of molecular biology that life comes from life, and origin-of-life research is nowhere close to proving this tenet wrong. The claim that the question of "whether life could have originated by chemical processes involving nonbiological components" has been resolved is simply not supported by the scientific data. The fact that evolutionists would make such a claim says more about their judgment than the state of the scientific research.

Huxley's Challenge

It appears that we should not look to Darwinists for a balanced assessment of the evidence. They consistently overrepresent the case for their theory. They present plenty of evidence but fail to explore the various interpretations of the evidence. In every case the evidence is ambiguous or even detrimental to evolution, but Darwinists tell a story of overwhelming success. We saw in chapters 2 and 3 that they fail to acknowledge the weight of the negative evidence. Now we see that they exaggerate the weight of what is supposed to be positive evidence.

The scientific argument against evolution is that this extraordinary theory lacks extraordinary evidence. Indeed, the positive interpretations that Darwinists apply to the scientific observations are easily countered when the whole story is carefully considered. It is not that there is no evidence for evolution—there is plenty. But the evidence brings along a variety of difficulties. Thomas Huxley's challenge that we set aside our preconceptions and, like a child, follow the scientific data is as relevant today as it was a century and a half ago.

6

Blind Guides

The Philosophical Argument against Evolution

From 1831 to 1836 Charles Darwin studied nature as he traveled around the world on the HMS *Beagle*. It was a crucial trip, for Darwin collected evidence that would later suggest to him that species had evolved in the wild. Perhaps the most famous of these findings came at the tiny Galapagos Islands, six hundred miles off the coast of South America.

On these islands there are more than a dozen finch species. And though they are all called finches, they have very different lifestyles. Some finches lived in coastal areas on the ground, others lived in forest trees, yet another lived in bushes. And the diet of these varieties varied considerably. One of the species ate buds and fruit, another prickly pear, others ate seeds, and others were insectivores. And one of the insectivores used a twig to fish out insects from crevices in the tree bark.

Darwin studied the wildlife on the Galapagos Islands, including the various finches. After the voyage Darwin considered what he had seen and the specimens that he and other members of the crew had brought back to England. After consulting other naturalists, it became apparent to Darwin that the different finches were indeed different species. The implications were enormous.

Today it may be difficult to understand why different finch species on the Galapagos Islands would be important evidence for Darwin. After all, we know the world is full of millions of different species. We have so many nature books and video documentaries showing all sorts of amazing creatures the world over. Biology seems to have no end of variety. Why are a few species of finch important?

Small-Scale Evolution

The reason the finches were important to Darwin is that the view of nature was far different in his day. In the centuries leading up Darwin's time, nature had been idealized as the harmonious and symmetrical creation of a wise and benevolent God. This idealization of nature was nothing new. The Pythagoreans in ancient Greece held that the sun, moon, planets, and stars were attached to rotating spheres that produced harmonious sounds, and Plato believed the universe was designed to be simple and elegant.

These notions continued two millennia later. For example, Johannes Kepler in the early seventeenth century sought to show how the Creator designed the heavenly motions according to musical harmonics. We shall see in chapter 7 that this sort of idealization was important in pre-Darwinian biology as well.

What is important for our purposes here is the strong acceptance of the idea that species must be fixed. *Unitas in omni specie ordinem ducit*—the invariability of species is the condition for order—was Linnaeus's pronouncement. Though Linnaeus softened this view to allow for a limited amount of change to occur, the notion that species must be fixed was closely associated with his well-accepted classification system. The nineteenth century began with the notion of *species as immutable* strongly in place.

A Break in a Dike

Darwin knew of a single finch species on the South American mainland, yet on the Galapagos Islands there were more than a dozen. And each one seemed appropriately adapted to its environment. Did the mainland finch immigrate to the islands and then adapt to the various environmental niches? This seemed a reasonable inference, but if so, it would, as Darwin wrote in his notebook, "undermine the stability of species."

In other words, the finches did not reveal to Darwin how fish could change to amphibia, or how amphibia could change to reptiles, or how reptiles could change to mammals. What the finches suggested was that species were not fixed—they could evolve. True, the amount of evolution was tiny, but the magnitude of the change was not so important. What was important was that the various types of finches were *different species*.

The doctrine of divine creation, as exemplified in the work of Linnaeus, had been dominant for centuries. But Darwin's finches seemed to be strong evidence against it. The Galapagos Islands presented an environment significantly different from the mainland. If God had independently created birds for each location, then they surely would be of a different sort. Why would the Creator make nearly identical species for such different environments? The better explanation was that the mainland finch immigrated to the islands and then radiatively adapted to its various environments. The one species became many.

It now seemed that the concept of species as fixed was in question and, along with it, the old doctrine of divine creation. We should not underestimate the significance of this shift. As evolutionist Ernst Mayr wrote: "The fixed, essentialistic species was the fortress to be stormed and destroyed; once this had been accomplished, evolutionary thinking rushed through the breach like a flood through a break in a dike."[1]

Farmers Cannot Be Creationists

This historical context is critical if we are to understand how evolutionists view the evidence. Darwinists have always hailed the evidence of small-scale evolution, as exemplified by Darwin's finches, as hard evidence for Darwin's theory. Other examples such as moths changing color, pesticide resistance, and HIV (the human immunodeficiency virus) are all touted as virtual proofs of evolution. Darwinists hold this evidence high, despite the significant problems we saw in chapters 4 and 5. In fact, there is no evidence that small-scale change is capable of the extraordinary large-scale evolution Darwinism requires.

But for evolutionists these problems are the subject of future research. They are undaunted, not because there is any hope of solving the problems, but because the evidence argues against divine creation. Problems with the scientific evidence become less important when there are no alternative theories. If divine creation is false, then evolution must be true. In this sense evolution is not limited to a particular naturalistic explanation, such as natural selection acting on random biological variation. Instead, evolution is simply *any* naturalistic explanation. Evolution is anything except divine creation. This is the reason evolution is such a flexible theory.

We know, for example, that insects gain resistance to pesticides, so evolutionist Jonathan Weiner rhetorically asks: "How can you be a creationist farmer any more?"[2] Resistance to antibiotics and pesticides do very little to prove evolution, but surely they prove the species are not fixed.

Whereas today we see the biological world as a complex web devoid of any simple, unifying principle, in Darwin's day it was still strongly idealized. Clearly there has been a dramatic shift in our view of biology, and Darwinism, more than anything else, brought about this shift. The high concept of species was an important part of the pre-Darwinian idealization, so evidence against the concept was evidence *for* the shift to Darwinism.

Comparative Anatomy

Another important concept in the pre-Darwinian idealization was the design of the species. The wise and benevolent Creator was expected to endow the species with the best possible design. As eighteenth-century naturalist Griffith Hughes put it, creatures he observed were "without defect, without superfluity, exactly fitted and enabled to answer the various purposes of their Creator, to minister to the delight and service of man, and to contribute to the beauty and harmony of the universal system."[3]

Much the same sentiment came from the eminent William Paley. His famous "watchmaker" argument was based on the exquisite designs found in biology. When one examines a watch it is obvious that the watch was created. So much more so with the body, which is ever more complicated than the watch.

But if the intricate design of the body revealed a divine Creator, what do nature's awkward designs suggest? Paley's work was required reading for university students in Darwin's day. The young Darwin had read Paley and was impressed. But Darwin would later conclude that species did not always appear to be "exactly fitted" for their purposes.

Inexplicable on the Theory of Creation

This became a powerful argument for Darwin. Indeed, though he admitted that evolution could only "to a certain extent"[4] account for comparative anatomy, he also argued that this evidence alone was sufficient to advance his theory.[5] The reason was, and remains today, that this evidence reveals Paley's sentimentalized version of divine creation to be false. Darwin repeatedly challenged the old doctrine on this evidence. Nature was, as he liked to say, "inexplicable on the theory of creation."[6]

For example, in the "theory of creation" that Darwin refers to, the species were supposed to have been uniquely designed. Common patterns found amongst different species reflected God's plan. Perhaps they were required for good engineering design. But some naturalists, such as Darwin, wondered if these repeated designs were really optimal. Wouldn't different functions require radically different designs? Nature seemed to make use of the same design over and over, even though the need was different. Creation was not fulfilling the naturalist's expectations, and for Darwin this was very curious:

> What can be more curious than that the hand of a man, formed for grasping, that of a mole for digging, the leg of the horse, the paddle of the porpoise, and the wing of the bat, should all be constructed on the same pattern and should include similar bones, in the same relative positions? How curious it is, to give a subordinate though striking instance, that the hindfeet of the kangaroo—which are so well fitted for bounding over the open plains—those of the climbing, leaf-eating koala—equally well fitted for grasping the branches of trees—those of the ground-dwelling, insect- or root-eating, bandicoots, and those of some other Australian marsupials should all be constructed on the same extraordinary type, namely with the bones of the second and third digits extremely slender and enveloped within the same skin, so that they appear like a single toe furnished with two claws. Notwithstanding this similarity of pattern, it is obvious that the hind feet of these several animals are used for as widely different purposes as it is possible to conceive. The case is rendered all the more striking by the American opossums, which follow nearly the same habits of life as some of their Australian relatives, having feet constructed on the ordinary plan.[7]

In this example Darwin points out that there are common patterns amongst different species—common patterns that do not seem to reflect good engineering design. Darwin could not explain how such organisms evolved with any level of scientific detail, but this could be investigated by later researchers. The strength of the argument lies in its implicit rebuke of divine creation, for why would God have designed such a mundane world? Why would God use the same pattern for different uses and, as Darwin observed, sometimes even in the same species?

> How inexplicable are the cases of serial homologies on the ordinary view of creation! Why should the brain be enclosed in a box composed of such

69

numerous and such extraordinarily shaped pieces of bone, apparently representing vertebrae? . . . Why should similar bones have been created to form the wing and the leg of a bat, used as they are for such totally different purposes, namely flying and walking? Why should one crustacean, which has an extremely complex mouth formed of many parts, consequently always have fewer legs; or conversely, those with many legs have simpler mouths? Why should the sepals, petals, stamens, and pistils in each flower, though fitted for such distinct purposes, be all constructed on the same pattern?[8]

Though Darwin did not know how the design of the crustacean or the flower could have been improved, he believed there must have been a better way and that God should have used it. God, according to Darwin, would not have made the brain or the bat that we find in nature, though he had little idea about how they actually worked.

Darwin's arguments for evolution are not open to scientific debate, for they rely on personal religious beliefs. One could just as easily argue that the Creator used the patterns found in homologous structures so that scientists could more easily analyze his creations and figure out how biology works. Or one could argue that the imperfections of nature that the homologies reveal are a manifestation of the burden of sin upon the world.

A Bungling Piece of Work

Darwin's claims about what the Creator can and cannot do are at the heart of evolution. Darwin's writings on evolution remain seminal today because he set forth the manner in which nature is to be interpreted.

Thus, in 1888 evolutionist Joseph Le Conte wrote extensively on how comparative anatomy reveals evolution because it refutes divine creation. For example, he wrote that the development patterns in fish revealed "a bungling piece of work" and that therefore they could not have been created.[9] In 1923 evolutionist H. H. Lane argued that "on the basis of special creation [homologies] have no meaning or else seem to limit the exercise of creative power."[10] And in 1952 evolutionist Arthur W. Lindsey wrote:

If special creation were the source of the many kinds of living things, it is reasonable to suppose that each would have the best possible equipment for its mode of life. Instead organisms often have adaptations which definitely resemble those of other species living under very different conditions. One structure may clearly show the essentials of the other, modified to serve a different use. They cannot logically be supposed to have been wholly independent in origin.[11]

These are not hand-picked examples. Darwin provided the religious interpretation, and it is used virtually every time arguments are made for evolution. I have never seen a forceful exposition of evolution that did not rely on these sorts of personal religious beliefs. Today's evolutionists rely on them no less than did earlier Darwinists. As Stephen Jay Gould put it, "odd arrangements and funny solutions are the proof of evolution."[12] Gould's point is not that evolution predicts such things, but rather that God would not have created them. It is a religious argument.

Mark Ridley, for example, asks "if whales originated independently of other tetrapods, should whales use bones that are adapted for limb articulation in order to support the reproductive organs? If they were truly independent, some other support would be used."[13] Of course, it is evolutionary conjecture that the whale bones are adapted for limb articulation, but Ridley's main point is that God would not have created whales like this.

Darwin's work was full of religious claims, and they remain crucial for today's evolutionists. The theory of evolution is true not because species obviously evolved from each other but because of the failure to reconcile God and nature. Darwin studied orchids in detail and again found underlying patterns. The orchids seemed to have been made of spare parts rather than individually created. For Darwin and modern evolutionists this argues for evolution because it argues against the possibility of divine creation. Gould sums up the argument as follows:

> Orchids manufacture their intricate devices from the common components of ordinary flowers, parts usually fitted for very different functions. If God had designed a beautiful machine to reflect his wisdom and power, surely he would not have used a collection of parts generally fashioned for other purposes. Orchids were not made by an ideal engineer; they are jury-rigged from a limited set of available components. Thus, they must have evolved from ordinary flowers.[14]

Notice how easy it is to go from a religious premise to a scientific-sounding conclusion. The theory of evolution is confirmed not by a successful prediction but by the argument that God would never do such a thing.

Vestigial Organs

Evolutionists hail similarities between species as great evidence for their theory. This argument becomes particularly forceful when the result is a useless structure or organ. The human appendix, for example, was long thought to be a useless organ, a vestige of the evolutionary process.

For evolutionists this is solid evidence for evolution, not because evolution predicted vestigial organs such as the appendix, but because such organs must not have been designed.

As De Beer put it, the existence of such useless structures "is inexplicable" except on the view of evolution.[15] Or as Lane explained, imperfect adaptation is "a hard blow to the advocates of special creation, for it would indicate a lack of skill or foresight not to be thought of in an all-wise and all-powerful Creator."[16]

For evolutionists, fitness is the universal design criterion. It is to be applied with equal force to the doctrine of creation as well as evolution. Apparently, God's purpose is primarily one of servicing creation. We shall examine evolution's concept of God in chapter 7. What is important here is simply that evolutionists rely on personal religious assumptions.

Of course, there is nothing wrong with proclaiming one's religious beliefs. The problem is not that evolutionists rely on particular religious assumptions; the problem is that evolutionists claim to have found a universal truth free of religious assumptions. Evolution, they say, is good science and should be taught in our public schools as such. Thus there is an internal contradiction within evolution. On the one hand it claims the higher ground of objectivity and science, but on the other hand it relies on particular and controversial religious claims that hardly have universal acceptance.

Few would disagree that science needs to make assumptions about the world, such as uniformity and simplicity. Experiments must be assumed to be repeatable, and simple explanations must be preferred to those with unnecessary details tacked on. But evolution's proclamations about what God can and cannot do go far beyond these time-tested assumptions. And of course, those proclamations have no basis in Scripture.

Pseudogenes

The word *pseudogene* is used to describe a segment of DNA whose sequence resembles a gene but is not used as normal genes are. It is not known what their function is, or even if they have a function. When DNA sequence data were first beginning to be compiled, evolutionists claimed that pseudogenes had no function. Evolutionists say pseudogenes are the result of mutations and that these disabled genes are passed down via common descent. This, evolutionists say, is why the same pseudogenes are found among similar species—they were inherited from a common ancestor. According to evolutionists, pseudogenes are powerful evidence for evolution.

For something that is supposed to be so convincing, however, there are quite a few problems with the pseudogene evidence. To begin with, evolution does not require pseudogenes. If there were no such thing as pseudogenes, evolutionists would be just as convinced their theory is true. Furthermore, pseudogenes do not always align as they should. Instead of pseudogenes revealing a pattern of common descent, they may be scattered about. The confusion, they say, may arise from the disappearance of a pseudogene, or perhaps by the independent creation of a pseudogene in different species. In other words, hypothetical evolutionary mechanisms can be used to explain any pattern.

Another problem with this evidence is that some pseudogenes apparently do have a function. What if functions are found for more pseudogenes? Will evolution's evidence reduce as our knowledge increases? As we saw in chapter 4, evolutionists simply change their story when this happens. In the case of vestigial organs, the evolution story was adapted as functions were discovered for them. And so it is with pseudogenes. Those that appear to have a function are given the appropriate evolutionary explanation—they are pseudogenes that regained function.

Finally, the idea that pseudogenes arose and were passed on does not require evolution. Yes, the pseudogenes of species have been inherited from their ancestors, but this does not mean we must resort to the unlikely story of evolution to explain them.

The point is that this evidence does not make evolution compelling. Pseudogenes are not powerful evidence for evolution from a scientific perspective. But there has always been another dimension to the pseudogene evidence. For example, evolutionists point to the L-gulono-gamma-lactone oxidase pseudogene. L-gulono-gamma-lactone oxidase is used in the synthesis of vitamin C and a disabled version of its gene appears in several species, including bats, the guinea pig, and primates, including humans.

Because we cannot synthesize vitamin C, we must obtain it in our diet, otherwise the terrible sickness known as scurvy is the eventual result. If the species were created, why would this important gene be disabled? Evolutionists also point out that the pseudogene is very similar in the different primate species in which it appears. Evolutionists say this is obviously the result of evolution. For why would a gene that doesn't work be designed into similar species? Clearly, they argue, the pseudogene was inherited via common descent. But the similarity of the pseudogene among those primates proves little. There are multitudes of similarities between the primates that evolutionists could use as evidence.

As pointed out above, if the evidence depends on the alignment of pseudogenes in an obvious pattern of common descent, it is not very strong. But this is not the evolutionist's point. The power of the argument is in its rejection of creation. It is not that evolution is the obvious explanation, but rather that design is obviously not the explanation. Evolutionist Kenneth R. Miller says pseudogenes must have evolved, for otherwise they would reveal a designer who "made serious errors, wasting millions of bases of DNA on a blueprint full of junk and scribbles."[17] Likewise for Edward E. Max pseudogenes are errors that "cannot reasonably be interpreted as having been 'designed.'"[18] These are powerful theological arguments which will be addressed in later chapters. For now, that is the point—they are *theological* arguments.

The Universal Genetic Code

Explicit references to God are easy to find in the evolution literature. But it is also typical for evolutionists to leave the assumption unspoken. For example, the universal genetic code, or DNA code, is claimed to be strong evidence for evolution. Mark Ridley explains that if the species were created, they wouldn't share the same code.[19] But Ridley's blunt religious claim is often made implicitly, though just as effectively.

The DNA code is used to read the information stored in the cell's genetic library, and essentially the same code is found in all species. Discovered in the second half of the twentieth century, the DNA code is hailed triumphantly by evolutionists as a great confirmation of Darwin's theory.

But we may ask, Why does the DNA code confirm Darwinism? The code and its attendant molecular machinery reveal a profound level of complexity about which Darwinism can only speculate. How could such a complex system have evolved? A great variety of explanations of the code's supposed evolution are currently under consideration. And they are filled with more speculation than hard fact.

The explanations call for the code to have evolved over time. For example, perhaps fewer amino acids were originally coded for, or perhaps the code distinguished between classes of amino acids rather than specific amino acids. Perhaps the code coevolved with the invention of biosynthetic pathways. Perhaps the alphabet was originally binary, or perhaps the words were only two letters long. Perhaps the original machinery was imprecise, so that a given gene did not always code for the same protein.[20]

And if the code could have evolved over time, then of course it is easily conceivable that it could have evolved into several different codes. In other words, evolutionary theory could explain the existence of multiple codes in nature. As such, evolution does not require there to be a single DNA code.

The universal genetic code is not required by evolution, nor can evolution explain how it evolved with any level of detail or certainty. And the code reveals an incredibly complex set of molecular machinery. It would seem that this hardly makes for good evidence. The reason evolutionists tout the DNA code as evidence, however, has nothing to do with these issues. Darwinists see the DNA code as evidence because they believe that if God had created the species he would have supplied them with *different* codes. The DNA code is evidence for evolution because it is evidence against divine creation.

Ridley completes the proof by informing us that God would never have used the same code for all species. But this premise often goes unstated. The universal genetic code is strong evidence for evolution not by any scientific argument, for the scientific evidence is arrayed against such a conclusion. The universal genetic code is strong evidence for evolution because it argues against creation.

Genetic Similarities

Sometimes the metaphysical assumptions behind evolution are in the form of claims of exclusivity. Scientific theories typically make certain predictions. If the prediction is verified, the theory could still be wrong, for there could be other explanations. But evolutionists sometimes say that only their theory can explain an observation. If *only* evolution is capable of explaining the observation, then divine creation cannot explain the observation. In other words, evolutionists are stating what God can and cannot do.

For example, evolutionist Carl Zimmer claims that genes that control the development of the nervous system in a wide range of organisms, including vertebrates and arthropods, must have evolved. The genes across this wide range of species are very similar. In fact, a gene from a fly can be inserted into a frog embryo and it will successfully do its job. "Such similar genes," Zimmer writes, "doing such similar jobs *must have* a common ancestry."[21] According to Zimmer, there is no other explanation.

Evolutionist Sean Carroll discusses another such gene-transplant experiment and arrives at the same strong conclusion. "There's *only one* inescapable conclusion," pronounces Carroll, "which is: If all of these branches have these genes, then you have to go to the base of that, which is the last common ancestor of all animals, and you deduce it must have had these genes."[22] Carroll claims that his interpretation is the only possible interpretation. The discussion may sound scientific, but ultimately it relies on assumptions about reality that are not open to scientific debate. There is no way, according to Carroll, that God could have created the genes this way.

In their evolution textbook Stearns and Hoekstra write that similar genes found in different species "establish shared descent from common ancestors."[23] How is it that Stearns and Hoekstra could possibly know that having similar genes implies common descent? This is the reverse of the scientific method. It is one thing to argue for one's theory; it is quite another to make an exclusive truth claim.

As we saw in chapter 5, the molecular clock is a hypothesis that leaves many questions unanswered. But from another perspective the molecular clock becomes powerful evidence. As evolutionist Thomas Jukes wrote, the molecular data are *"only* comprehensible within an evolutionary framework."[24] In other words, aside from evolution there are *no other* explanations for the data.

How can evolutionists such as Jukes, Stearns, Hoekstra, Zimmer, and Carroll so unequivocally claim that genetic similarities establish evolution? How is it that common descent is the "only one inescapable conclusion?" This can be true only if divine creation cannot explain the data. Evolutionists often make high truth claims not characteristic of science, and such claims are always an indicator of nonscientific premises.

The Hierarchical Pattern

Nature's species fall into a striking pattern of hierarchical clusters. The species tend to cluster into groups, but those groups also fall into yet larger groups, and so on. This is not direct evidence for evolution since it can predict a variety of patterns. But from Darwin to today, evolutionists have repeatedly used this as evidence against creation. Darwin began the tradition with this blunt religious claim: "The several subordinate groups in any class cannot be ranked in a single file, but seem clustered round points, and these round other points, and so on in almost endless cycles. If species had been independently created, no explanation would have been possible of this kind of classification."[25]

Evolutionist George Carter admitted that the pattern does not "necessarily imply evolution." But "if species are separately created," explained Carter, "there is no reason why they should be created in large groups of fundamentally similar structure."[26]

Likewise, Joel Cracraft writes that under the idea of creation, the "similarities observed among organisms cannot be shared so as to produce a hierarchical pattern of groups within groups."[27] Niles Eldredge concurs:

"Could the single artisan, who has no one but himself from whom to steal designs, possibly be the explanation for why the Creator fashioned life in a hierarchical fashion—why, for example, reptiles, amphibians, mammals, and birds all share the same limb structure?"[28]

For evolutionists the conclusion is obvious. The evidence forbids divine creation—at least the kind of creation that evolutionists have in view. Darwin and the others could not conceive of a God who would have his species so grouped, so evolution must be true.

The Same Phylogeny (Evolutionary Tree) from Different Design Features

We saw in chapter 5 that this evidence is not as convincing as evolutionists claim. But ever since Darwin evolutionists have interpreted this evidence in a way that converts it into a much stronger argument for evolution. To understand their interpretation we need to briefly consider the work of Richard Owen, a leading British anatomist and contemporary of Darwin. In Darwin's day, as we shall see in chapter 7, the belief that God is better viewed as transcendent and not involved in the details of creation was popular in Victorian England. So it is not surprising that there were at least some thinkers, such as Owen, who were considering the idea that God created the species indirectly, via some sort of directed evolutionary process.

Owen was attempting to fit the species into a divine plan. Owen envisioned what he called archetypes—common plans which God used when creating the species. God would create and later refine according to the archetype. But Owen's idea made God appear to be more of a tinkerer than wise Creator. Owen's strained mixture of God and evolution was the perfect foil. Darwin rebuked Owen in *Origin* and Huxley attacked him mercilessly in public.

A key question which both Owen and Darwin were trying to answer was how to explain what seemed to be unnecessary similarity in biology. For example, why were there similar designs where the function was different? Owen's answer was that they represented the Creator's archetypes. Darwin's answer was that they were the leftovers of the unguided process of evolution. Where Owen saw order, Darwin saw happenstance.

Owen's idea of archetypes was reasonable but his mixing of God and evolution was not. His entire system was left vulnerable and Darwin called it hopeless. Better to view such similarities as the result of random effects in a contingent process rather than the plan of a tinkering creator. With the triumph of Darwinism this view is now canonized. The concept of archetypes is considered to be firmly disproved and evolutionists too easily conclude that similarities between species come from chance not reason.

The result is that rather mundane findings are now hailed as stunning confirmations of evolution. Early blood immunity studies revealed that humans are more closely related to apes than to fish or rabbits. Evolutionists viewed the similarity as unnecessary and therefore a great confirming evidence for their theory. This is a case where the underlying metaphysics were subtle but, as we shall see in chapter 11, they would ultimately become explicit.

Fossil Evidence

In chapter 4 we saw that the fossil evidence raises many questions about evolution. Nonetheless, Darwinists have long maintained that the fossils prove evolution. About fifty years ago the evolutionist George G. Simpson noted how puzzling the fossil record could appear.[29] Yet Simpson wrote elsewhere that "there really is no point nowadays in continuing to collect and to study fossils simply to determine whether or not evolution is a fact. The question has been decisively answered in the affirmative."[30]

Extinctions and the Rise of Complexity

How can evolutionists be so persuaded by the fossil evidence? The answer, as revealed in the writings of evolutionists, is that the fossils disprove divine creation. It may be a mystery how one species could evolve from another, but the progression of complexity in the fossil record speaks conclusively, wrote H. H. Lane, "against the traditional view of creation."[31] Or as de Beer put it, unless "one is prepared to believe in successive acts of creation and successive catastrophes resulting in their obliteration, there is already a strong presumptive indication that evolution has occurred."[32]

The fossil record is characterized by stasis and the rapid appearance of new forms. When similar forms are arranged, there are always ambiguities. In some cases there are too many species, leaving evolutionists with a multitude of possible lineages and the need for explanatory devices such as convergent evolution. Some species may overlap in time and show no sign of merging one into the other. Or there may be unique and advanced forms appearing too early.

But the fossil record most definitely reveals a progression of higher forms accompanied by the extinctions of the older forms. For evolutionists such as Kenneth R. Miller, this requires evolution, for otherwise we are left with a designer who "just can't get it right the first time. Nothing he designs is able to make it over the long term."[33] For Douglas Futuyma the regularity of the fossil record accords with evolution, not creation.[34]

Futuyma's God would never create the species according to any such sort of order or regularity. Berra also finds this trend a problem for creation because advanced organisms usually appear later than the primitive ones:

> This sequential appearance of different groups at different times, the more advanced appearing, in general, later than the more primitive, is predicted by evolutionary theory. It cannot be reconciled with creationism.[35]

Mark Ridley finds that fish, amphibians, reptiles, and mammals would not appear in that order in the fossil record if they had been separately created.[36] And for Stephen Jay Gould, God would not have created species in these sorts of sequences, and therefore evolution must be true: "What alternative can we suggest to evolution? Would God—for some inscrutable reason, or merely to test our faith—create five species, one after the other . . . to mimic a continuous trend of evolutionary change?"[37]

Gould's suggestion that God would need an inscrutable reason for mimicking evolution simply is not supported by the scientific data. As we saw in chapters 4 and 5, the evidences, including the fossil record, do not point to evolution. But evolutionists have specific ideas about how God must act. For Gould, God simply would not have created the fossil species we find, so they must have evolved.

There is also the problem that fossil species sometimes are similar to the living species in the same region. "Why," asks Miller, "should such a unique set of animals be found in exactly the same place as their closest fossil relatives?" Surely God would not create similar species in the same locale. "There could be just one answer," states Miller. The species were not created; instead, "a process of descent with modification linked the present to the past."[38]

Fossil Anomalies

But if the fossil record evinces too much order, it also has an arbitrary aspect that, for some evolutionists, does not accord with creation. For the fossil species do not always seem to progress in any particular direction. "What could have possessed the Creator," asks Futuyma, "to bestow two horns on the African rhinoceroses and only one on the Indian species? "[39] Miller also finds evidence against the Creator in the elephant fossils. There are, explains Miller, dozens of elephant or elephantlike fossil species dating back as much as 50 million years. Trends in the design of the trunks and tusks can be found amongst these species. Using these trends, the species can be compared, classified, and even arranged in an evolutionary tree if one believes in evolution. And we should believe in evolution, according to Miller, for can we possibly believe there is a Creator behind this haphazard arrangement?

> This designer has been busy! And what a stickler for repetitive work! Although no fossil of the Indian elephant has been found that is older than 1 million years, in just the last 4 million years no fewer than nine members of its genus, *Elephas*, have come and gone. We are asked to believe that each one of these species bears no relation to the next, except in the mind of that unnamed designer whose motivation and imagination are beyond our ability to fathom. Nonetheless, the first time he designed an organism sufficiently similar to the Indian elephant to be placed in the same genus was just 4 million years ago—*Elephas ekorensis*. Then, in rapid succession, he designed ten (count 'em!) different *Elephas* species, giving up work only when he had completed *Elephas maximus*, the sole surviving species.[40]

Miller obviously has specific ideas of what the designer is and is not allowed to do. First off, the designer must be sensible to us, going about his work as we see fit. Repetitive work seems unlikely—he certainly wouldn't go about making ten different *Elephas* species in rapid succession. In fact, tallying up all the millions of different species ever found, the Creator must have been constantly at work, and this too, for Miller, is hard to believe.[41]

An Internal Contradiction

In chapters 2 through 5 we saw that evolution failed the test of science. The negative evidence against evolution is daunting, and the so-called positive evidence does little to help the situation. Many wonder where Darwinists gain their confidence in the light of such difficulties. The answer is that Darwinists view the empirical evidence through a particular theological lens. The paltry evidence is converted into unbeatable arguments when a particular religious filter is applied.

It is not that the data alone strongly suggest evolution—they don't. But the data can be used against divine creation if one's belief about creation is suitably crafted. Darwinists rely on a particular notion about God and creation in order to argue that he couldn't have made this world. And given that creation is not possible, then evolution is the only other alternative. As the Dutch theologian Herman Bavinck, an early and perceptive critic, observed:

> Darwin was led to his agnostic naturalism as much by the misery which he observed in the world as by the facts which scientific investigation brought under his notice. There was too much strife and injustice in the world for him to believe in providence and a predetermined goal. A world so full of cruelty and pain he could not reconcile with the omniscience, the omnipotence, the goodness of God. . . . The discovery of the so-called law of "natural selection" brought him accordingly a real feeling of relief, for by it he escaped the necessity of assuming a conscious plan and purpose in creation. Whether God existed or not, in either case he was blameless. The immutable laws of nature, imperfect in all their operations, bore the blame for everything.[42]

This is why evolution's only real constraint is that it be restricted to naturalistic explanations. Beyond this, evolution is not constrained to any particular explanation about the creation of the species. Evolution is not defined by what it is; rather, it is defined by what it isn't.

It has been said that evolution relies on an underlying assumption of naturalism or even atheism. But naturalism is the effect, not the cause. Darwinists promote the nonbiblical concept of a God who must be distanced from the world. Therefore, only naturalistic explanations will do. Naturalism is a corollary, not a premise, of Darwinism.

As Darwin put it, "If I have erred in giving to natural selection great power . . . or in having exaggerated its power . . . I have at least as I hope, done good service in aiding to overthrow the dogma of separate creations."[43] God must be distanced from the world—beyond that evolution can be quite flexible. The result is that evolutionists may be willing to part ways with natural selection but not with the idea that the process of evolution is naturalistic and unguided.

The philosophical argument against evolution is that it contains an internal contradiction. Darwinists claim religion plays no role in their theory, but religion lies at its very foundation. It is the constant thread running throughout Darwinism. Evolutionists reject a particular religious explanation, but in doing so they proclaim their own religion. Declaring what or how God may not create is just as religious as declaring what or how he does create.

7

Another Gospel

The Theological Argument against Evolution

When Princeton theologian Charles Hodge critiqued Darwin's theory of evolution, he noted that Darwin made plenty of references to God. Hodge wasn't sure quite what to make of Darwin's concept of God. Darwin did everything he could to steer his theory away from God, yet he referred to God repeatedly in his arguments for evolution. What sort of God did Darwin have in mind?

Concepts of God vary greatly. What is interesting is that these concepts are not strictly the purview of believers. It seems that one need not be one of the faithful to hold strong convictions about God. Skeptics often seem to be equally committed to their own ideas about God.

In the Middle Ages theologians produced a variety of arguments for the existence of God. One of the most intriguing of those theistic proofs was Anselm's (1033–1109) ontological proof. Anselm began with the premise that God is the greatest thing we can conceive. In his peculiar language, God is "that than which no greater can be conceived."[1] If there is something greater than what we are thinking of, we are not thinking of God. Surely it goes without saying that a God who exists is greater than one who does not. So the skeptic's claim that God does not exist is invalid, because he is confusing another notion for God. When he says that God does not exist, he is not really thinking of God at all.

Skeptics cannot help but feel that something is not quite right about Anselm's proof. It engaged great thinkers for centuries, but it seems altogether too easy. Though Anselm's proof may have little influence today, there is an underlying message that is useful for our purposes. Perhaps Anselm didn't prove God's existence as he had intended, but he did make an important point about how people tend to think of God. His argument calls attention to the tendency toward a trite deity. Too often people talk of God when they have in mind something much less. In fact, the evolutionist's definition of God almost always goes unstated and must be inferred. If we refer to God, Anselm reminds us, we ought to own up to our word. At the very least, we should define what we mean.

In the Image of Man

Darwin often referred to the God of the Bible in arguing for evolution, but his view of God was not consistent with the Scriptures. According to the Bible, there is much that can be known about God, both from his creation and his revelation in Scripture. But there is also much that cannot be known. "His greatness no one can fathom," writes the psalmist, and "Such knowledge is too wonderful for me."[2]

We, the created beings, cannot know everything about God, as he reminds us through his prophet: "'For my thoughts are not your thoughts, neither are your ways my ways,' declares the LORD. 'As the heavens are higher than the earth, so are my ways higher than your ways and my thoughts than your thoughts.'"[3]

Nonetheless, there is a human tendency to speak for God without respecting our mortal limitations. This is not outright disbelief but instead is a more subtle form of skepticism. Rather than arguing against God altogether, people often simply assume a lesser God. Indeed, doctrines of God that Anselm would have found wanting have been popular in the church for centuries.

In the Scriptures Job and his friends speculated about God's purposes. Then God descended upon them in a storm. "Who is this," the Lord asked, "that darkens my counsel with words without knowledge?"[4] In this frightening passage God showed his anger and revealed the absurdity of man cross-examining God: "Will the one who contends with the Almighty correct him? . . . Would you condemn me to justify yourself?"[5] In the end Job was humbled: "Surely I spoke of things I did not understand, things too wonderful for me to know."[6]

Scripture does not give us a picture of a God who is to be analyzed by man. Yet in the centuries leading up to Darwin, just such a tradition grew in the church. The deists were obvious and extreme examples of subjecting God to man's judgment.

Deism

God's revelation in the Bible is not always intuitive. For example, God is one, yet there is a Trinity: God the Father, God the Son, and God the Holy Ghost. English deism was a seventeenth- and early-eighteenth-century movement that sought an intuitive explanation of the Bible story that anyone could understand.

According to deists, the true religion should be something that anyone could figure out. It should be deducible from nature, rather than having to be revealed by God. Instead of God's revelation being given at a particular time and place in history and then spreading throughout the world, it should be always available to all people in all places and at all times.

With deism, Christianity became less historical. The recorded words of biblical figures at specific points in history became less important. The evidence for God was in creation, not revelation, and therefore it wasn't necessary for God to be active in the world. In deism, God was reduced to a distant memory. He created the world long ago and has since been uninvolved with the world. The Bible and its God were losing their authority, and humanity was being simultaneously elevated.

This was John Toland's (1670–1722) message in his 1696 work, *Christianity not Mysterious*. Toland made human reason the ultimate guide, and there was little place for a personal, immanent God in such a system. In Toland's pantheistic view the deity was identical to the universe itself. Ten years later Matthew Tindal (1655–1733) wrote the bible of deism, *Christianity as Old as the Creation*. The point of the title is that the only portions of Christianity that are worthwhile, or perhaps legitimate, are those that derive from reason; and reason has been available to us since the beginning. Any so-called revelations that are beyond reason, such as humanity's fall and God's saving atonement, are superstitions.[7]

The deists had definite opinions about how God would proceed if he were to reveal himself to humanity. Instead of God having to save the world, he had to make the world in such a way that man could save himself.

Though deism was not a lasting movement, it was influential. Its notions were apparent in liberal Anglicanism as well as on the continent and in America. John Tillotson (1630–1694), archbishop of Canterbury, was a contemporary example.[8] In later years, Voltaire (1694–1778), Hermann Samuel Reimarus[9] (1694–1768), Thomas Paine (1737–1809), Thomas Jefferson, and Darwin's own grandfather, Erasmus Darwin, would number among those influenced by deism.

Implicit in deism was the assumption that the fall of humanity was not so very damaging as the Bible portrays it to be. Although the Scriptures explained that the effect of the fall corrupted our very thinking, the deists believed that humanity could reason its way to God. Instead of rebelling against God, deism saw humanity deducing God from nature. Human reason was hardly seen to be in a fallen state.

Of course, this diluted version of the fall did not begin with the deists, and in the eighteenth century it was hardly peculiar to the deists. Thinkers at most points in the spectrum, even those who opposed the deists, held to a similar view of the fall. And this view naturally fed the view of a lesser God. One didn't need to be a deist to conclude that if humanity is not really fallen, and we can reason our way to faith in God, then God need not be the active, sovereign God portrayed in the Scriptures.

Joseph Butler

More than anyone else, the influential Anglican bishop Joseph Butler was credited with defeating deism. But Butler's approach was not unlike the deists' emphasis on reason and their dilute version of the fall. Butler differed with the deists more in their evaluation of the evidence than with their underlying assumptions.

Butler attacked deism and defended Christianity. He pointed out deism's internal contradictions. The deists used one standard of proof for the Scriptures but a lower standard for their own deistic beliefs. In defense of Christianity, he argued that the evidence of the created world makes it probable. Furthermore, there is utility in Christianity. It makes us happy. Given its probability and utility, Butler argued that it is more prudent to believe than not to believe.

Butler defeated the deists, but in the process he cast Christianity more as a mathematical problem than a living faith. We should believe not so much because God has saved us but because it behooves us to believe. Faith was the logical choice.

It is not that Butler explicitly advocated a remote or lesser God. But his approach of computing probabilities about God could have the effect of lowering God. The Creator had made the world in such a way that Christianity could be deduced. God was not needed to soften the rebellious heart, because the heart wasn't all that rebellious to begin with.

Natural Theology

The idea that God should create according to our reason and intuition was also evident in the version of natural theology that was popular in the seventeenth and eighteenth centuries. A variety of clergy and scientists alike wrote prolifically on how the naturalist can find God in creation, but their concept of God was often less informed by Scripture than it was by human sensibilities.

It is often said the natural theologians tried to prove the existence of God. More often than not, however, their aim was simply to extol God's wonderful world and how it reflected his divine power and wisdom. And while the world certainly does contain incredible creations that point to the Almighty, it also has its evils.

If one is going to make claims about what the world reveals, one must consider all the evidence. Yet the natural theologians focused mostly on the pleasing aspects of the world. This was the message of men such as John Ray (1627–1705), William Derham (1657–1735), and George Cheyne (1671–1743). Their message was that most things in the world revealed purpose and design and that the world was designed to maximize happiness. The natural theologian could look at practically any situation in nature and find it to be precisely arranged so as to produce the most happiness for men and animals.[10]

The evil in the world could be viewed in a variety of ways. It could simply be ignored, or it could be said to be ultimately a good thing. Pain, for example, reminds us just how good pleasure really is. John Ray, on the other hand, ascribed nature's apparent inefficiencies to the notion of an intermediate creative force called "plastic nature."

This concept came from the Cambridge Platonists earlier in the sixteenth century. The idea was that God would not, as Ray put it, "set his own hand as it were to every work, and immediately do all the meanest and trifling'st things himself drudgingly, without making use of any inferior or subordinate Minister."[11] The subordinate minister or agent was plastic nature, which, unlike the Creator, was not infallible or irresistible. Instead, plastic nature had to contend with the ineptitude of matter. The result was those "errors and bungles" of nature.[12] This pious account explained natural evil and allowed Ray to focus on the glorious aspects of creation.

Historians often erroneously refer to the natural theologians as orthodox in their religious beliefs. No doubt these men espoused a great God. But this alone does not make for Christian orthodoxy. The Scriptures also speak of God's sovereignty, providence, and creation acts. Scripture leaves little room for a distant God who requires a plastic nature, or for downplaying the evils of the world. As one historian noted, "those interested in the wisdom of God in creation were so enamoured with what they saw that they practically subsumed the concern with revelation."[13]

The Bible describes a glorious creation that God formed according to his good pleasure. But the Scriptures also explain that creation, in the words of the apostle Paul, was subjected to frustration: "For the creation was subjected to frustration, not by its own choice, but by the will of the one who subjected it, in hope that the creation itself will be liberated from its bondage to decay and brought into the glorious freedom of the children of God. We know that the whole creation has been groaning as in the pains of childbirth right up to the present time."[14]

The Scripture's explanation of creation is complicated. Creation is awesome but also groaning and in bondage to decay. With eighteenth-century natural theology, however, the evils of the world could be only awkwardly explained. The reason was that natural theology focused on humanity's reasoned examination of creation. Humanity was supposed to be objective, and creation was supposed to be pleasantly arranged.

William Paley

Natural theology reached a pinnacle in the work of William Paley (1743–1805). He used examples of biological design brilliantly, and his argument remains cogent today. But along with the other natural theologians of his time, Paley struggled to explain the evil in the world.

Earlier natural theologians such as Ray and Derham were important sources for Paley, so it is no surprise that Paley also emphasized the happiness of nature's creatures as evidence for God. Paley even presented a proof for his claim that God "wills and wishes the happiness of his creatures."[15]

Paley began with the premise that when God created human beings he either wished for their happiness, wished for their misery, or was indifferent. Paley then ruled out God wishing for our misery, for if that were the case, he didn't do a very good job. God certainly could have made things much more miserable. Likewise, if God were indifferent, then why do our senses have the capacity to receive pleasure, and why is there such an abundance of external objects fitted to produce it?

By the process of elimination, Paley thought he had proved that God wills and wishes the happiness of his creatures. In a few short pages, and without reference to Scripture, Paley reduced a profound theological question to a triviality. The implication, Paley concluded, was that "the method of coming at the will of God, concerning any action, by the light of nature, is to inquire into the tendency of that action to promote or diminish the general happiness."[16]

It wasn't always clear whether Paley found evidence for God in nature or whether he interpreted nature according to his concept of God. In any case, Paley presented a decidedly happy view of nature:

> It is a happy world after all. The air, the earth, the water, teem with delighted existence. In a spring noon, or a summer evening, on whichever side I turn my eyes, myriads of happy beings crowd upon my view. The insect youth are on the wing. Swarms of newborn flies are trying their pinions in the air. Their sportive motions, their wanton mazes, their gratuitous activity, their continual change of place without use or purpose, testify their joy, and the exultation which they feel in their lately discovered faculties. A bee amongst the flowers in spring, is one of the most cheerful objects that can be looked upon. Its life appears to be all enjoyment; so busy, and so pleased: yet it is only a specimen of insect life.[17]

The passage continues at length as Paley exults in the bliss of nature. As with other natural theologians, Paley is sometimes portrayed as an orthodox Christian. But Paley overemphasized certain aspects of the faith. The deviation of Paley's version of God from the scriptural version is sometimes subtle. Paley was certainly a strong believer, and no doubt the Scripture does present a loving God. Furthermore, Paley did not ignore the problem of evil.

But Paley's theodicy did not seriously deal with the evil in the world or with the scriptural picture of a sovereign God. Paley wrote that the many different natural laws, which normally work for good, sometimes inadvertently "thwart and cross one another" and so cause natural evil.[18] It sounded more like deism than orthodox Christianity. Where the Scriptures explained that God can cause calamity, Paley saw calamity as the occasional inadvertent glitch of an otherwise finely tuned machine.[19]

And Paley's claim that God wills and wishes the happiness of his creatures also seems to reflect a limited view of God. Scripture does affirm that God loves his creation and that he brings his people to a place of abundance.[20] But the way is not always easy.[21] We need look no further than the struggles of the apostles and, indeed, the Lord's crucifixion. One need not be a Christian to know that there is more to God's salvation plan than the maximization of our earthly happiness.

"It is strange," wrote one historian, that the natural theologians, "could turn such a blind eye to the negative aspects of existence. This has certainly tended to throw a shadow of ridicule over all their arguments."[22]

The Bridgewater Treatises

Paley's work was highly influential. In the nineteenth century Paley was required reading for young students such as Darwin. And though Paley's approach was not entirely without controversy, it did have a lasting influence well into Darwin's time.

One outstanding example was the Bridgewater Treatises, which were a highly visible sample of natural theology in the Victorian era. According to the will of the eighth earl of Bridgewater, who died in 1829, the treatises were to demonstrate "the Power, Wisdom, and Goodness of God, as manifested in the Creation."[23]

Eight eminent scientists contributed to the treatises, and in many of the passages the message was that the pleasant aspects of God could be seen in creation. It is not that the gruesome aspects of nature were completely ignored, but when they were considered, they were force-fitted into this happier view of God.

One author, for example, translated nature's bloodshed and death, which, after all, was relatively swift and painless, into a divine "dispensation of benevolence."[24] He argued that modern science revealed the "infinite wisdom and power and goodness of the Creator."[25] Another author was able to find signs of a happy God in even the most extreme examples of parasitic behavior.[26]

In the centuries leading up to Darwin, natural theology was popular in England, but it often had in view a rather two-dimensional Creator. In this version of natural theology, it seemed that God was often defined according to human sensibilities rather than according to Scripture. These natural theologians rightly pointed to the obvious signs of God in the creation and to his "infinite wisdom and power and goodness." But they had difficulty with the evils of creation. Too often they tried to explain those evils by falling back to the design argument. Rather than use scriptural explanations, they force-fit evil into their paradigm as some sort of design trade-off.

A Framework for Darwin

Natural theology was lopsided. Yes, the world is amazing, but the natural theologian's happy view of nature could hardly be justified in light of the real world. It is easy to see why this version of natural theology supplied a ready source of material for its opponents.

Some opponents were religious skeptics, but often the debate was between believers. The alternative typically offered to natural theology's good God who designed the world was a transcendent God who was utterly unconcerned with the world's evils.

With this transcendent God, the world was created not by direct divine action but by secondary causes—God's natural laws in operation. Such a God was all the wiser. Instead of dabbling in his creation, this removed and uninvolved God had designed a self-sufficient creation. His natural laws did the work automatically.

This uninvolved God must have great power and wisdom. These positive attributes provided opponents of natural theology with a divine sanction—they too could point to a good God. As the Anglican cleric Thomas Burnet wrote in the seventeenth century: "We think him a better Artist that makes a Clock that strikes regularly at every hour from the Springs and Wheels which he puts in the work, than he that hath so made his Clock that he must put his finger to it every hour to make it strike."[27] In other words, it is better for God to create a self-sufficient machine than to make one needing divine intervention.

In a sense this God was not so different from the natural theologian's God. Paley and the others had argued that God's wisdom and power was evident in creation, and these virtues were required all the more in the transcendent God. Indeed, Paley had little difficulty with this view of God. He argued that a watch requires a designer, but he also agreed that a self-replicating watch simply reinforces the argument.

Suppose we were to find a watch, wrote Paley, that "possessed the unexpected property of producing, in the course of its movement, another watch like itself." This should increase our admiration of the contrivance and our "conviction of the consummate skill of the contriver."[28] The idea of a strictly transcendent God was not so much a rejection but a modification of the eighteenth-century natural theology.

The concept of a removed God was nothing new. The ancient Gnostics believed in such a God. For them spirit was good, matter was evil, and the two must be utterly separate. Gnosticism had some influence in the early church, and less radical forms of Gnosticism have been popular throughout the age of the church. But Gnosticism selects Scripture carefully. Scripture's transcendent but involved God who created and governs the world is nowhere to be found among Gnostic influences.

David Hume

One of the most well-known attacks on natural theology came in David Hume's (1711–1776) *Dialogues concerning Natural Religion,* published posthumously in 1779. Hume wrote an engaging dialogue among his three fictitious characters, Cleanthes, Demea, and Philo, to make his argument for a transcendent God.

Cleanthes, who represented Butler and the natural theologians such as Ray, argued that the world was a happy place and that this was evidence for God. Hume had an easy time ridiculing this view. Philo, who represented Hume, and Demea agreed that "a perpetual war is kindled amongst all living creatures" and that nature is so arranged so as "to embitter the life of every living being."[29]

If the natural theologian's God would have the world be happy, and yet the world is not happy, then we must seek a different God. Philo concludes, against Cleanthes, that God does not will the happiness of man or animal.

Cleanthes's design argument was powerful. Philo admits the argument was a great challenge for him, but it is neutralized by the evil in the world. "I needed all my skeptical and metaphysical subtlety to elude your grasp," admits Philo, but "here I triumph."[30]

Philo's solution was to make God more mystical. He charged Cleanthes with anthropomorphizing God. Cleanthes, said Philo, made God out to be too much like his human creatures. For example, the natural theologians were fond of comparing the human body with machines such as clocks. No one doubts that a clock was designed, so why not the body as well? Hume, through his character Philo, used the problem of evil to negate this argument. Better to view God as distant and unknowable.

Hume had no problem with God being infinitely powerful and wise, but Hume's God was to be solely transcendent and incomprehensible. For Hume, God is a God of faith, not the God of history:

> But as all perfection is entirely relative, we ought never to imagine that we comprehend the attributes of this divine being, or to suppose that his perfections have any analogy or likeness to the perfections of a human creature. Wisdom, thought, design, knowledge; these we justly ascribe to him; because these words are honorable among men, and we have no other language or other conceptions by which we can express our adoration of him. But let us beware, lest we think that our ideas anywise cor-

respond to his perfections, or that his attributes have any resemblance to these qualities among men. He is infinitely superior to our limited view and comprehension; and is more the object of worship in the temple, than of disputation in the schools.[31]

In other words, while we may have faith in God, we must not think we understand him well enough to infer his actions in the material world. He may be the God of our hearts, but not of our heads. But Hume's argument is one-sided. He charges the natural theologians with anthropomorphizing God when they look at the design in the world. But it seems that it is Hume who is guilty of this when he looks at the evil in the world. The natural theologian's core argument—that complexity requires a designer—seems far less heroic than Hume's argument that the world's obvious evil is evidence for a God who did not create the world.

Victorian England

Hume's ideas on how to view God were part of a growing sentiment. By Darwin's day the Victorian elite were increasingly subscribing to a noninterventionist conception of God. Miracles were once viewed as part of God's providence, but now were viewed as clumsy. The lord chancellor of England, Henry Peter Brougham (1778–1868), said they proved nothing but the exercise of miraculous power, and they left the Creator's trustworthiness in question.[32]

The Reverend William Conybeare claimed that "the Bible is exclusively the history of the dealings of God towards men."[33] His point was that the Bible was spiritual and should not be used to infer the history of creation. God was an object of faith but not part of natural history. Likewise, the Reverend Baden Powell called for God to be limited to things religious. He wrote in 1838:

> Scientific and revealed truth are of essentially different natures, and if we attempt to combine and unite them, we are attempting to unite things of a kind which cannot be consolidated, and shall infallibly injure both. In a word, in physical science we must keep strictly to physical induction and demonstration; in religious inquiry, to moral proof, but never confound the two together. When we follow observation and inductive reasoning, our inquiries lead us to science. When we obey the authority of the Divine Word, we are not led to science but to faith. The mistake consists in confounding these two distinct objects together; and imagining that we are pursuing science when we introduce the authority of revelation. They cannot be combined without losing the distinctive character of both.[34]

The message here is that religion and science are to be kept separate. God is retained to supply the former, but it would never do to consider him in the latter. The Creator is used to explain morality but is simply disconnected from the physical world.

In geology Charles Lyell argued that only natural processes should be used to explain the earth's history. Though there was plenty of evidence that catastrophic events had taken place in the distant past, Lyell urged uniformitarianism. God did not form the world directly; it evolved via natural laws.

In 1850, a decade before Darwin went public with his theory of evolution, John Millais's painting *Christ in the House of his Parents* was first exhibited at the Royal Academy. In the painting, the boy Jesus had injured his hand in his father's carpentry shop. Mother Mary attended to the boy while father Joseph continued with his work. The scene was realistic, with wood scraps lying all about and workers going about their duties. But the scene was altogether too realistic for the Victorians, who sought a more distant God.

The Scriptures said that God became flesh and dwelt among us.[35] He knew sorrow, pain, temptation, and joy. But this view of God was lost on the Victorians. They emphasized God's wisdom, power, and transcendence. Could he really have bruised his hand in a messy carpenter's shop? The *Times* complained that the painting was revolting, for its "attempt to associate the holy family with the meanest details of a carpenter's shop, with no conceivable omission of misery, of dirt, even of disease, all finished with the same loathsome meticulousness, is disgusting." *Blackwood's Magazine* said, "We can hardly imagine anything more ugly, graceless and unpleasant," and Charles Dickens called the painting "mean, odious, revolting and repulsive."[36]

These are but a few examples of the growing neo-Gnosticism in Darwin's time. The Victorians could no more believe than the Gnostics could that Christ the Savior could become involved with creation. "If Christ is to be taken seriously as the Savior," explained a historian about Gnostic beliefs, "how can he actually be part and parcel of this material cosmos?"[37]

The Gnostics could not believe God became a man for the same reasons that they could not believe God directly created the world—they could not envision God involved in a world so fraught with misery. Similarly, just as the Victorians were troubled by Millais's depiction of the human side of Jesus, they also would have trouble with the idea that God created the biological world, apparently so full of inefficiencies, anomalies, and useless bloodshed.

Darwin's God

We saw in chapter 6 that beginning with Darwin, evolutionists have relied on religious arguments about how God would have created the world. Darwin's version of God is the foundation of his theory of evolution, just as concepts of God were foundational for the deists, natural theologians, and neo-Gnostics. And Darwin's version of God is clearly evident in these earlier traditions.

The eighteenth-century natural theologians emphasized God's wisdom and goodness. Therefore, the world should be a good and happy place. Darwin incorporated this one-sided view of God to argue that the species must have evolved. Over and over he pointed to the predation and apparent inefficiency in nature to support his claim that it must have arisen independent of divine direction.

Darwin had no better explanation for complexity than had Hume. But as with Hume, Darwin assumed that nature's designs required that God be remote. And also like Hume, Darwin turned to a mystical version of God.

For example, Darwin made an argument against natural theology that paralleled Hume's warning against anthropomorphizing God. Darwin pointed out that while it is tempting to see God as the master engineer who crafted complex organs such as the eye, this would make God too much like man. Darwin agreed that the perfection of the eye reminds us of the telescope, which resulted from the highest of human intellect. Was it not right to conclude that the eye was also the product of a great intellect? This may seem the obvious answer, but Darwin warned against it, for we should not "assume that the Creator works by intellectual powers like those of man."[38] Better to imagine the eye as the result of natural selection's perfecting powers than have God work as man does.

Since Darwin, evolutionists have continued to rely on this trite concept of God. They openly proclaim what God would and would not do in his acts of creation. They speak of what they think is reasonable and sensible for God to do, apparently unaware that, aside from personal religious belief, there is no justification for their claims. They use the word *God*, though they intend something much less than an orthodox Christian understanding of God. For example, taken together, the species of the world form a striking hierarchical pattern. Darwin claimed that "if species had been independently created, no explanation would have been possible of this kind of classification."[39]

How could Darwin possibly justify this bold theological claim? What Scripture was available to support this incredible assertion? The answer is, of course, none. Instead, Darwin had assumed the lofty position of judge, pronouncing just what God could and couldn't do.

Evolutionists since Darwin

We saw many similar examples in chapter 6. How could H. H. Lane support his assertion that homologies "have no meaning" on the basis of special creation? And how could Arthur W. Lindsey support his claims that it is "reasonable to suppose" that God would not have created the species as we find them?

Where did Mark Ridley learn that the DNA code would not be shared amongst different species if they were created? And how can he justify his claim that God was created?[40] How does Carl Zimmer know that similar genes doing similar jobs must have a common ancestry?

Where did Kenneth Miller learn that God would not have created the different elephantlike species? "This designer has been busy," Miller taunts, "and what a stickler for repetitive work!"[41] Miller is confident of his knowledge of God, but where did it come from? What private Scripture is informing Miller of these great truths about God?

Likewise Douglas Futuyma believes in a highly restricted God. Here are three paragraphs quoted in sequence, but not continuous in the original:

> If God had equipped very different organisms for similar ways of life, there is no reason why He should not have provided them with identical structures, but in fact the similarities are always superficial.[42]

> The facts of embryology, the study of development, also make little sense except in the light of evolution. Why should species that ultimately develop adaptations for utterly different ways of life be nearly indistinguishable in their early stages? How does God's plan for humans and sharks require them to have almost identical embryos?[43]

> Take any major group of animals, and the poverty of imagination that must be ascribed to a Creator becomes evident.[44]

Like Darwin, Futuyma sets strict limits for his God. Certainly, God must not offend our sensibilities. As Stephen Jay Gould put it: "Odd arrangements and funny solutions are the proof of evolution—paths that a sensible God would never tread but that a natural process, constrained by history, follows perforce. No one understood this better than Darwin. Ernst Mayr has shown how Darwin, in defending evolution, consistently turned to organic parts and geographic distributions that make the least sense."[45]

Without these religious arguments evolution has only weak evidence to fall back on. Evolution began with, and continues to rely on, a very peculiar concept of God. Indeed for many, their god is really no god at all, but a mere idol.

The theological argument against evolution is that its theological assertions fail Anselm's test. Evolutionists use the word *God*, but it is merely a term of convenience. Evolutionists are not seriously considering an Absolute God. Instead they appeal to a simplistic, feel-good God; indeed, they require such a God to make their theory appear convincing.

8

A Reason for Hope

The Only Explanation for Life

Charles Darwin's theory of evolution is deeply flawed. As we have seen in chapters 2 through 7, it has profound problems. The Darwinists must swallow the camel of complexity and other negative evidence while straining at the gnat of positive evidence. Then, while making claims about God, they insist their theory is nothing more than objective science. And their god is really no god at all.

But if evolution is wrong, then what is right? How did life come about, and why is there so much evil and inefficiency in the world? Evolution provided an explanation for the evil in the world. Evil arose by the unguided workings of nature—the interplay of blind natural laws that have no goal in sight. Evolution failed to explain the wonders of the world. We can hardly believe that evolution's unguided forces somehow produced the most complex things we know of. But evolution provided a way of understanding this flawed world—natural forces were the culprit. Some way, somehow, the blind action of nature's laws and interactions caused what we now find to be so undesirable.

Come Let Us Reason

Darwinism does provide an explanation for natural evil, but it relies on the against-all-odds story that the species evolved. It is probably the most successful theodicy ever devised, so long as one ignores the underlying absurdity.

But we cannot overlook the absurdity. Evolution gives us a way of understanding evil, but it fails to explain the existence of life. What we need is an explanation for all the data. How is it that we have a world that is at once so incredibly complex and elegant, yet so profoundly evil? How can the most complex thing known in the universe—the human brain—contemplate iniquities beyond measure? On the one hand, the brain obviously could not have evolved. On the other hand, its action could not be the intentions of a designer. Both explanations fail.

The problem is that while we endlessly pursue naturalistic solutions that protect God from the world, we ignore the obvious solution. As I argued in *Darwin's God*, Darwinism is really all about God. God wouldn't have created this world, say the evolutionists. But Darwinists have a false god in mind. What about the real God?

A Sovereign Creator

We saw in chapters 6 and 7 that the eighteenth-century natural theologians as well as evolutionists had much to say about God, but that their God was rather trite. Let us now examine another possibility—the ancient Scriptures. What does the Bible say about our God, ourselves, and our world?

To begin with, consider evolution's argument that species are sometimes not well adapted. While evolution can accommodate this imperfection, it is not to be expected if the species were divinely created. As evolutionist Arthur W. Lindsey wrote, if the species were created "it is reasonable to suppose that each [species] would have the best possible equipment for its mode of life."[1]

Hence, adaptation becomes the universal design criterion. Of course adaptation is supposed to be the result of evolution via natural selection. But evolutionists apply it equally to the Creator. This view is not peculiar to Lindsey or even to evolutionists. Gottfried Leibniz had argued that God optimized the good-to-evil ratio in the world. God was the master craftsman, working according to universal design criteria. And the pre-Darwinian natural theologians believed God's goal was to make the world happy.

Before going further we should mention that the adaptations of the species to their surroundings are indeed amazing. It is one of the wonders of biology how all the various species seem to fit so well into their niches.

But occasionally there are design aspects of species that don't seem quite right to evolutionists. Are the awkward tail of the peacock and the pentadactyl pattern signs of divine wisdom? "When we compare the anatomies of various plants or animals," evolutionist Douglas Futuyma pronounces, "we find similarities and differences where we should least expect a Creator to have supplied them."[2]

But where, we may rightly ask, did Futuyma and the others learn about God? For the Scriptures give no indication of such a creator. Consider God's revelation to Job on how he created the ostrich:

> The wings of the ostrich flap joyfully,
> but they cannot compare with the pinions and feathers of the stork.
> She lays her eggs on the ground
> and lets them warm in the sand,
> unmindful that a foot may crush them,
> that some wild animal may trample them.
> She treats her young harshly, as if they were not hers;
> she cares not that her labor was in vain,
> for God did not endow her with wisdom
> or give her a share of good sense.
> Yet when she spreads her feathers to run,
> she laughs at horse and rider.[3]

For four chapters God makes it clear that he is sovereign and creates according to his good pleasure. The stars and constellations that come forth in their seasons are God's creations. The lioness hunts her prey, the donkey pays no attention to the driver's shout, the ox has great strength, the horse laughs at fear, and the hawk spreads its wings and takes flight by God's wisdom.[4]

"Do you give the horse his strength or clothe his neck with a flowing mane?" God asks Job. And "does the eagle soar at your command and build his nest on high? He dwells on a cliff and stays there at night; a rocky crag is his stronghold. From there he seeks out his food; his eyes detect it from afar. His young ones feast on blood, and where the slain are, there is he."[5]

There is a natural temptation to rationalize the world—to make our own sense of God's creation. But Scripture describes God as sovereign. He does not create according to some optimization formula that we can derive. Yes, God created the species perfectly, but he did so according to his own good pleasure. Scripture teaches that God created all things by his will.[6] He also created the things unseen; that is, the spiritual realm of existence. Creation is distinct from God but dependent on God. He is both transcendent and immanent—greater than creation yet within creation.

"As the heavens are higher than the earth," says the Lord, "so are my ways higher than your ways and my thoughts than your thoughts."[7] But this awesome God is also involved with creation. "My Father is always at his work to this very day," said Jesus, "and I, too, am working."[8] Indeed, God knows every bird in the mountains and the creatures of the field are His.[9]

The world is God's doing. He designed it and he created it—he is above the world and in the world. Scripture does not give us a picture of a creator who is limited or constrained by certain aspects of creation. Nothing is outside of his control. He does not create according to pre-existing natural laws—he created the laws as well as the matter that obeys those laws.

All this is to say that God is sovereign over the world and can create the species as he wishes. God is not limited in his creation acts to our idealized notions of perfection. When Darwin told the Christians that homologies are inexplicable "on the ordinary view of creation" he was relying on a man-made version of God that is nowhere to be found in the Scriptures.[10]

A Fallen World

God's thoughts are not our thoughts, and he is sovereign in his creation acts. Hence we should not expect to immediately understand all his works. If we find that the ostrich lacks wisdom, we should not take this as evidence that it wasn't created by God.

The Bible also speaks of an altogether different reason why we may not understand the workings of nature—it is corrupt. Humanity's rebellion against God did not simply bring about a human separation from God. Original sin did not have only moral implications—it also brought about a dramatic change to the natural order.

For example, sin caused the ground to be cursed.[11] After the fall the ground would produce thorns and thistles, and by the sweat of man's brow would the land be fruitful. Even more significant is the death brought about by sin. As Paul wrote, "sin entered the world through one man, and death through sin, and in this way death came to all men, because all sinned."[12] Obviously, sin had a dramatic effect on the natural order.

The Glory of God

The Scriptures describe a fallen creation, but they also explain that creation still reveals God's glory. After each stage of creation God said it was good, and after the final stage he said it was "very good." In contrast to the Gnostics who claimed matter was evil and spirit good, the Bible affirms the originally innate goodness of the material world God has created.[13]

And in this goodness we can see God's glory. "The heavens declare the glory of God," wrote King David, "the skies proclaim the work of his hands."[14] "The whole earth is full of his glory," wrote Isaiah.[15] "God made the earth by his power," wrote Jeremiah, "he founded the world by his wisdom and stretched out the heavens by his understanding."[16]

Other biblical passages give extended descriptions of how creation reveals God's power and glory. In the final chapters of the book of Job, God makes it abundantly clear to Job and his friends that the wonders of creation are his. "Where were you," the Lord asks Job, "when I laid the earth's foundation?" "Have you ever given orders to the morning?" or "Have you seen the gates of the shadow of death?"[17]

Or again, Psalm 104 extols God's glory as revealed in the world. "Praise the LORD, O my soul," it begins, for the Lord is "very great."[18] The psalmist proclaims that God "stretches out the heavens like a tent and lays the beams of his upper chambers on their waters,"[19] and that "He set the earth on its foundations; it can never be moved."[20]

God "makes springs pour water into the ravines"[21] for all the beasts of the field and "makes grass grow for the cattle, and plants for man to cultivate."[22] God has made the "wine that gladdens the heart of man, oil to make his face shine, and bread that sustains his heart."[23]

"The moon marks off the seasons, and the sun knows when to go down."[24] God brings darkness, and then the beasts of the forest prowl. The sun rises, and they steal away. Then man goes out to his work, to his labor until evening. All these and more display the power and wisdom of the Creator:

> How many are your works, O LORD!
> In wisdom you made them all;
> the earth is full of your creatures.
> There is the sea, vast and spacious,

> teeming with creatures beyond number—
> living things both large and small.
> There the ships go to and fro,
> and the leviathan, which you formed to frolic there.
>
> These all look to you
> to give them their food at the proper time.
> When you give it to them,
> they gather it up;
> when you open your hand,
> they are satisfied with good things.
> When you hide your face,
> they are terrified;
> when you take away their breath,
> they die and return to the dust.
> When you send your Spirit,
> they are created,
> and you renew the face of the earth.
>
> May the glory of the LORD endure forever;
> may the LORD rejoice in his works—
> he who looks at the earth, and it trembles,
> who touches the mountains, and they smoke.[25]

Over and over the Scriptures portray a glorious creation and a sovereign Creator. And they make a connection between humanity's choice to sin and the natural order. Unlike evolution, which attempts to explain the quirks and evils of nature but cannot explain complexity, and unlike the design argument, which accounts for complexity but cannot explain evil, the Scriptures give us a complete picture of the world. The Bible predicts both evil and complexity. God has created a glorious and awesome world. On the one hand, it includes complexity beyond measure, but on the other hand, it includes predation. And it has been corrupted by humanity's fall from grace.

Our Sinful Condition and the Fall

In addition to corrupting the natural world, the fall has also darkened our thinking. Humanity is far from God and far from the truth. Instead of thinking clearly about things, humanity will produce deceptive philosophies. We are to expect deceptions in the form of fine-sounding arguments that appeal to human sensibilities and traditions.[26]

The Scriptures warn that though knowledge will increase, "none of the wicked will understand."[27] People will be "always learning but never able to acknowledge the truth."[28] They will "not put up with sound doctrine. Instead, to suit their own desires, they will gather around them a great number of teachers to say what their itching ears want to hear. They will turn their ears away from the truth and turn aside to myths."[29]

The Scriptures explain that humanity will contrive its own myths in its rejection of the truth. One of the biggest such myths of our day is evolution. The Lord's warning through the prophet Isaiah certainly seems relevant to evolutionary thinking:

> Woe to him who quarrels with his Maker,
> to him who is but a potsherd among the potsherds on the ground.
> Does the clay say to the potter,
> "What are you making?"
> Does your work say,
> "He has no hands"?
>
> Woe to him who says to his father,
> "What have you begotten?"
> or to his mother,
> "What have you brought to birth?"
>
> This is what the LORD says—
> the Holy One of Israel, and its Maker:
> Concerning things to come,
> do you question me about my children,
> or give me orders about the work of my hands?
> It is I who made the earth
> and created mankind upon it.
> My own hands stretched out the heavens;
> I marshaled their starry hosts.[30]

Yes, fallen humanity does give orders to God about his creative works. The strong arguments for evolution are theological in nature, stating what God would and would not do in creating the species. In addition to this, the very idea of God creating the species is unacceptable to fallen humanity. The problem, say evolutionists, is that one cannot test the theory of creation. God can do anything, so any biological question can be resolved by simply saying that God did it that way. It seems that God's great power now works against him. God can do as he wishes with creation, so he must be rejected. As Darwin made clear, what was important was not his particular theory of evolution but rather the rejection of creation: "Whether the naturalist believes in the views given by

Lamarck, by Geoffroy St. Hilaire, by the author of the 'Vestiges,' by Mr. Wallace or by myself, signifies extremely little in comparison with the admission that species have descended from other species, and have not been created immutable: for he who admits this as a great truth has a wide field open to him for further inquiry."[31] Or as Darwin's friend J. D. Hooker put it, if theories of divine creation are "admitted as truths, why there is an end of the whole matter, and it is no use hoping ever to get any rational explanation of origin or dispersion of species—so I hate them."[32]

Whereas earlier scientists took no issue with the idea that science was a tool for investigating God's creation, Darwin now argued that such thinking was outside of science: "On the ordinary view of the independent creation of each being, we can only say that so it is;—that it has pleased the Creator to construct all the animals and plants in each great class on a uniform plan; but this is not a scientific explanation."[33]

Thirty years later Joseph Le Conte argued that the origins of new species, though obscure and even inexplicable, must have a natural cause. To doubt this, he warned, is to doubt "the validity of reason, and the rational constitution of organic Nature." For Le Conte divine creation was not rational. Evolution, he triumphantly concluded, "is as certain as the law of gravitation. Nay, it is far more certain."[34]

Today this notion continues. John Rennie, editor-in-chief of the venerable *Scientific American*, recently wrote that "rather than expanding scientific inquiry," the design arguments "shut it down." For how, Rennie rhetorically asks, "does one disprove the existence of omnipotent intelligences?" Rennie continues: "For instance, when and how did a designing intelligence intervene in life's history? By creating the first DNA? The first cell? The first human? Was every species designed, or just a few early ones?"[35]

Since we cannot answer these questions we must not consider God in science. For Niles Eldredge, the key responsibility of science—to predict—becomes impossible when a sovereign Creator is entertained: "But the Creator obviously could have fashioned each species in any way imaginable. There is no basis for us to make predictions about what we should find when we study animals and plants if we accept the basic creationist position. . . . the creator could have fashioned each organ system or physiological process (such as digestion) in whatever fashion the Creator pleased."[36]

This idea was also important for Paul Moody: "Most modern biologists," he wrote, "do not find this explanation [that God created the species] satisfying. For one thing, it is really not an explanation at all; it amounts to saying, 'Things are this way because they are this way.'

104

Furthermore, it removes the subject from scientific inquiry. One can do no more than speculate as to why the Creator chose to follow one pattern in creating diverse animals rather than to use differing patterns."[37]

And Tim Berra warns that we must not be led astray by the apparent design in biological systems, for it "is not the sudden brainstorm of a creator, but an expression of the operation of impersonal natural laws, of water seeking its level. An appeal to a supernatural explanation is unscientific and unnecessary—and certain to stifle intellectual curiosity and leave important questions unasked and unanswered."[38] In fact, "Creationism has no explanatory powers, no application for future investigation, no way to advance knowledge, no way to lead to new discoveries. As far as science is concerned, creationism is a sterile concept."[39]

For their part, Darwinists have never been shy about using religion to promote their ideas. They have told us that God must not have created this world, and hence it must have evolved. If we want evidence of humanity giving orders to God, we need look no further than Darwinism.

Furthermore, Darwinists use the idea that divine creation is stifling and goes against reason. Here we see Darwinism in perhaps its rawest form. The Creator is rejected outright, for it would never do to allow the world to be created. The science may not always work quite right, but only natural causes must be allowed. The very idea of God creating the world is not allowed, for humanity must have its own way.

Obviously this sentiment makes no sense—we cannot on logical grounds simply define God out of the picture. This is a hijacking of science, a blunt example of humanity's rejection of God. Science, once the instrument for exploring creation, now is allied with skepticism. Believe if you want, but keep such beliefs private, for we must not entertain the God hypothesis in science.

Science was built on the shoulders of Christian thought, and a great number of scientists are Christians, but now in the guise of neutrality, science rejects God outright. It is, however, anything but neutral. The idea that God ought not be considered in scientific investigations entails the assumption that God *need not* be considered in scientific investigations. That is, science can make good progress and describe the world accurately without reference to God. If God created the world, he could have done so only via natural laws. There must be no direct divine intervention.

This is hardly a neutral stance. Historically it comes from the idea that God must be distanced from creation. God is transcendent but not immanent—greater than creation but not within creation. He is removed from history and relegated to the sphere of faith. He is rejected from the issues of the day. The secularization of everything from public policy and law to the media and education is the result of this tradition. We have today precisely what Hume urged more than two hundred years ago: the privatization of God. As Hume concluded, God is better viewed as "the object of worship in the temple, than of disputation in the schools."

The view is not neutral, for it presupposes that the world has an independent existence apart from God. It creates what the Dutch theologian Herman Bavinck called a chasm in the sphere of reality between God and the world. In short, it relies on a nonscriptural view of reality.

It is not that we do not know God, but that we reject him from practically all aspects of our lives. Instead we turn to our secularized idols. The apostle Paul wrote of such rejection in his letter to the Romans. "For although they knew God," Paul warned, "they neither glorified him as God nor gave thanks to him, but their thinking became futile and their foolish hearts were darkened."[40]

Rejection of God leads to futile thinking. One need not be a scientist to see the manifestation of the Creator in the creation. It is obvious to all, and people are without excuse, for they have rejected God and suppressed the truth in unrighteousness. Humanity's rebellion against God makes people blind to the obvious signs of the Creator, and ultimately foolish. "Although they claimed to be wise," wrote Paul, "they became fools." Instead of acknowledging God, humanity exchanged the truth for a lie and "worshiped and served created things rather than the Creator."[41]

Elsewhere Paul warned of hollow and deceptive philosophies—fine-sounding arguments that mislead.[42] "Where is the wise man? Where is the scholar? Where is the philosopher of this age?" asked Paul. For God had made foolish the wisdom of the world.[43] Evolution may have the world's approval—those who think they are wise and learned may give their assent—but it reveals the foolishness that results from humanity's rejection of God.

The creation abounds with evidence for the Creator, but it is ignored by the rebellious, fallen creature. Belief in God is not a scientific problem to be worked out; it is not an intellectual puzzle that can be investigated and researched. Rather, it is intertwined with the sin problem. We fail to believe, not because we haven't figured it out, but because we have rebelled. Lack of faith is not a head problem; it is a heart problem.

A Final Accounting

Humanity's fall and rejection of God are not abstract theological concepts. Sexual immorality, impurity, evil desires, greed, anger, rage, malice, slander, and filthy language are some of the more obvious signs of sin.[44] We are told that "people will be lovers of themselves, lovers of money, boastful, proud, abusive, disobedient to their parents, ungrateful, unholy, without love, unforgiving, slanderous, without self-control, brutal, not lovers of the good, treacherous, rash, conceited," and "lovers of pleasure rather than lovers of God."[45]

But sin need not be so obvious. The well-mannered as well as the profane may be equally committed to sin. As Jesus made clear, evil thoughts are also abominable. God hates not only murder and adultery, but also their forefathers, unrighteous anger and lust.[46]

Is sin not so very important in the eyes of God? Does God lower the bar for his creatures? Nowhere do the Scriptures give any such indication. "If your right eye causes you to sin," Jesus warned, "gouge it out and throw it away. It is better for you to lose one part of your body than for your whole body to be thrown into hell."[47] Indeed, rather than lowering the bar, Jesus instructs us to be "perfect, therefore, as your heavenly Father is perfect."[48]

Sin is real. It has had terrible consequences, and ultimately it must be dealt with. "Each of us," Paul warns, "will give an account of himself to God."[49]

Our sin is ever before us. Sin is not the occasional slip or mistake of an otherwise good person, but rather a repeated pattern.[50] In our honest moments we know that we have things to hide. We love what God hates. As King David confessed, "My sins have overtaken me, and I cannot see. They are more than the hairs of my head, and my heart fails within me."[51]

Though we may lie to ourselves and downplay our own sin, it cannot escape the view of God. He searches out the hearts[52] and knows all of our private thoughts. There is no variation or shadow of turning in God.[53] He knows all our evils, and he will not set them aside arbitrarily. Just as with the keeping of accounts in this world, so too there must be an accounting in the next world.

We cannot escape the long arm of the Lord. He created us, and to him we will return. Each of us will in turn die and face our Maker, who is a consuming fire.[54] Here on earth we so often place our hopes in material means rather than trusting in God. Instead of placing our faith in God, we turn to worldly schemes. But when we face God there will be no such choices. God will accuse the sinner, and there will be none to rescue.[55] He is a righteous judge who does not relent.[56]

God's Grace: Jesus Christ

The messages of human sinfulness and divine judgment are not easy. They are frightening for some, humiliating for others. But we have willingly entered into sin. We have created this awful situation. It would be foolish for us to reject the Bible's message because it is burdensome for us, for we ourselves have filled up our own burden.

The Bible consistently portrays a stark situation for us. Humanity, in its sinful state, cannot save itself. Nor did Jesus soften the picture any. Our sinful state is a stumbling block, and there is only one solution. The people asked Jesus, What are the works that God requires? Jesus answered, "The work of God is this: to believe in the one he has sent."[57] Jesus atoned for our sins, and those who believe are justified by the blood of Jesus.

This is the good news. In our stark situation there is a great hope. "For God so loved the world that he gave his one and only Son, that whoever believes in him shall not perish but have eternal life. For God did not send his Son into the world to condemn the world, but to save the world through him."[58]

This is good news in many ways. Not only are we saved from our own sins, but we are saved by the Judge himself. Our freedom does not depend on our own works. Human frailty and temptation cannot get in the way of our salvation. Instead, God himself sets us free. The blood of Jesus is our protection, for he took on the sins of the world. Because we have a free pardon, we need not argue a losing case.

Our work is to believe him who was sent—the Lord Jesus Christ. Of course, believing that Jesus received the punishment that was due us, that he was resurrected from death, and that he is Lord and Savior has great implications. Faith in Jesus brings about an overhaul in our attitude. It is not a one-time commitment but a lifetime commitment. We do not sign a card and then return to our old ways. The Scriptures explain that the believer is to repent, be baptized, and embrace a new life. We are to grow in our spiritual walk according to God's Word and in the fellowship of mature believers.

We now know the good news, and we rejoice with trembling.[59] There may be trials and tribulations, but we know that in all things God works for the good of those who love him.[60] Joseph's brothers tried to kill him, but what they intended as evil, God intended for good, for the saving of many lives.[61] Commit your way to the Lord and trust in him, and he will make your righteousness shine like the dawn.[62]

9

What Has Been Made

One Long Parable

When Jesus was crucified there were those in the crowd who gambled for his clothes and others who mocked and insulted the dying Savior. "You who are going to destroy the temple and build it in three days, save yourself," they jeered. "Come down from the cross, if you are the Son of God." The chief priests, teachers of the law, and the elders also mocked him. "He saved others," they said, "but he can't save himself! He's the King of Israel! Let him come down now from the cross, and we will believe in him."[1]

They were witnessing the most profound event in history, yet they expected something more. They so misunderstood God and his plan that the suffering Messiah made no sense to them. Why would a king so empty himself? If Jesus were the Son of God, they thought, he should have made a display of it then and there. Then they would believe.

Those witnesses had very particular expectations of God. The true spiritual meaning of Jesus' suffering and death was lost on them. They had witnessed so many of his miracles, and in three days he would rise from the dead, but they still would not believe.

Jesus had even told them a revealing parable of a beggar named Lazarus and a rich man to illustrate their situation. After the two men died, angels carried Lazarus to Abraham's side, and the rich man suffered in hell. The rich man called to Abraham and asked him to send Lazarus to warn his five brothers so they would not also be put in that place of torment. Abraham replied that his brothers have Moses and the Prophets; let them listen to them. "No, father Abraham," the rich man cried, "but if someone from the dead goes to them, they will repent." "If they do not listen to Moses and the Prophets," Abraham relied, "they will not be convinced even if someone rises from the dead."[2]

Jesus had performed an abundance of miracles. He had even brought dead people back to life. But none of this would sway the skeptics. Even when he rose from the grave they would not believe. The crucifixion was for them a justification for their rejection of Jesus. If he was the Christ, then why not display his powers at that moment?

What an incredible example of humanity misunderstanding God. But it is just an example; all humanity has made the same mistake in generation after generation. We are in no position to judge those witnesses of the crucifixion when we are just as guilty.

Humanity Misunderstands God

What lesson can we learn from this? What mistakes do we make in our generation when it comes to understanding God and his plan? Those skeptics witnessing the crucifixion may have been thinking of biblical prophecies foretelling God coming with power. They probably expected a conquering king rather than a suffering servant. But the Scriptures are full of prophecies that tell of the Messiah taking on our sins and bearing the punishment that we are due.

Centuries before, the prophet Isaiah had explained how the Messiah would come. He would be "despised and rejected by men." He would take up our infirmities and carry our sorrows, and yet we would consider him stricken by God.[3] Isaiah could not have been more blunt:

> But he was pierced for our transgressions,
> he was crushed for our iniquities;
> the punishment that brought us peace was upon him,
> and by his wounds we are healed.

We all, like sheep, have gone astray,
each of us has turned to his own way;
and the LORD has laid on him
the iniquity of us all.

He was oppressed and afflicted,
yet he did not open his mouth;
he was led like a lamb to the slaughter,
and as a sheep before her shearers is silent,
so he did not open his mouth.
By oppression and judgment he was taken away.
And who can speak of his descendants?
For he was cut off from the land of the living;
for the transgression of my people he was stricken.[4]

Those skeptical witnesses of the crucifixion were probably not think-ing of these verses. And they were probably not thinking of a psalm of David that foretold the very events surrounding them. Written a thou-sand years earlier, centuries before crucifixion was used as a means of execution, David's illustration of the scene is graphic, complete with the mockers and skeptics:

All who see me mock me;
they hurl insults, shaking their heads:
"He trusts in the LORD;
let the LORD rescue him.
Let him deliver him,
since he delights in him."
. . . .
I am poured out like water,
and all my bones are out of joint.
My heart has turned to wax;
it has melted away within me.
My strength is dried up like a potsherd,
and my tongue sticks to the roof of my mouth;
you lay me in the dust of death.
Dogs have surrounded me;
a band of evil men has encircled me,
they have pierced my hands and my feet.
I can count all my bones;
people stare and gloat over me.
They divide my garments among them
and cast lots for my clothing.[5]

The mockers and skeptics were a fulfillment of prophecy. Had they loved God's word more than their own opinions, then perhaps they would have understood the real meaning of the crucifixion. Perhaps they would have stood in wonder and awe rather than in contempt.

Therefore, a lesson we can learn is to be good students of God's Word. The apostle Paul wrote that "all Scripture is God-breathed and is useful for teaching, rebuking, correcting and training in righteousness."[6] We need to humble ourselves to God's Word, testing all things, even our own intuition, against it.

The Paradigm of Perfection

But like those skeptics at the crucifixion, modern thinkers have also imposed their own ideas about how God would work. When it comes to the study of origins, there has been a strong tendency to view God as the perfect craftsman. After all, the book of Genesis tells us that God created the world and that it was "very good."[7] And David wrote that the "heavens declare the glory of God; the skies proclaim the work of his hands."[8]

But the story is more complicated than this. Creation may be "very good," but Paul tells of how it was subjected to frustration. That it is in bondage to decay and is groaning as in the pains of childbirth.[9] And God tells Job of how the donkey does not hear the shouts of the driver and of how the ostrich treats her young harshly for he did not endow her with wisdom.[10]

The Scriptures plainly tell us not to expect a perfect world. God has not made the world to be always optimal in a material sense, and the fall of humanity brought about its own destructive forces. Nonetheless, optimality is just what many believers expect from God's creation. Gottfried Leibniz, for instance, acknowledged the evil in the world but assumed that it was there so as to optimize the good-to-evil ratio. Less evil would have meant much less good. Likewise, the cleric Thomas Malthus admitted that when populations grow faster than the food supply it produces some hardship, but he reasoned that it also "produces a great overbalance of good."[11]

Or, as we saw in chapter 7, there were the natural theologians before Darwin's time who described the world and everything in it as perfectly designed and arranged to work as a perfect machine. As Paley put it, God "wills and wishes the happiness of his creatures." Paley extolled creation as a perfect machine, for it was "a happy world after all. The air, the earth, the water, teem with delighted existence. In a spring noon, or a summer evening, on whichever side I turn my eyes, myriads of happy beings crowd upon my view."

Darwin inverted Paley's bright optimism, but not his assumptions about God:

> We behold the face of nature bright with gladness, we often see super-abundance of food; we do not see or we forget that the birds which are idly singing round us mostly live on insects or seeds, and are thus constantly destroying life; or we forget how largely these songsters, or their eggs, or their nestlings, are destroyed by birds and beasts of prey; we do not always bear in mind, that, though food may be now superabundant, it is not so at all seasons of each recurring year.[12]

Darwin's point of course, was not that Paley's view was nonscriptural, but rather that God's creative acts ought to be reconsidered. Darwin proposed a mechanistic creative process, not a different view of God. And with the acceptance of Darwinism, this view of God has been, in turn, all the more accepted. Evolutionists, as we saw in chapter 6, have continued to rely on this nonscriptural view of God to justify their theory.

So when evolutionist Ken Miller asks if God would "really want to take credit for the mosquito," he is echoing a long-standing tradition, not some biblical truth.[13] The idea that God would create only a perfect or optimal world, in a material sense, has been a dominant paradigm among naturalists and philosophers for centuries.

We might call this the *paradigm of perfection*. From Kepler and Leibniz to Paley and Darwin, the assumption that creation is an end unto itself and a thing that God would certainly strive to make perfect if it were possible, has driven our thinking in the historical sciences.

But this perspective has failed us. It has no basis in Scripture and has led to absurdity. We need to reconsider this view and consider replacing it with a historic doctrine that theologians have known about all along—the doctrine of general revelation.

General Revelation

The universe may have many purposes, but the church has always taught that, as God's creation, it complements God's Word. The Scriptures are called special revelation, and the world is called general revelation. As the apostle Paul wrote:

> The wrath of God is being revealed from heaven against all the godlessness and wickedness of men who suppress the truth by their wickedness,

since what may be known about God is plain to them, because God has made it plain to them. For since the creation of the world God's invisible qualities—his eternal power and divine nature—have been clearly seen, being understood from what has been made, so that men are without excuse.[14]

Paul explains that creation itself makes plain what may be known about God. And certainly throughout the Bible we find analogies in creation for spiritual truths. We are told that the blessed man is like a tree planted by streams of water, but the wicked are like chaff that blows away in the wind.[15] God's incredible providence is displayed by the beauty and intricacy of the lilies,[16] and the kingdom of heaven is like a little bit of yeast that works through a large amount of dough.[17] But false teachers are like a brood of vipers,[18] and the fool who comes back to them is like a dog that returns to its vomit.[19]

Did God have to search for these analogies? Was it serendipity that creation just happened to be full of analogies to spiritual truths that are given in the Scriptures? Or did God create the world to contain signs of his truth?

The eighteenth-century theologian Henry Drummond held the view that natural laws are analogous with spiritual laws. We can learn something about the spiritual world from the natural world. Before him, Johann Goethe claimed that transitory things are but parables. These are different ways of reflecting on the doctrine of general revelation. Nature has, as Herman Bavinck put it, "lain in the thought of God before it came into being."[20]

Creation, at least in part, contains revelation for us to consider. In other words, Scripture tells us not only that God created the world, but that the world serves to communicate some of his revelation. Perhaps the caterpillar, imprisoned in its cocoon and then emerging as a butterfly, is symbolic of the crucified and then resurrected Christ.

If creation serves as part of God's revelation, then why should we think it must be designed to be perfect or optimal in a material sense? Paley's premise that God "wills and wishes for our happiness" seems at odds with Scripture, which tells us to consider it "pure joy, my brothers, whenever you face trials of many kinds, because you know that the testing of your faith develops perseverance. Perseverance must finish its work so that you may be mature and complete, not lacking anything."[21]

We should view creation's purpose as communicating truths and contributing to our salvation, not as a gratification of our senses. Instead of considering the paradigm of perfection, we should consider the paradigm of revelation. Of course other factors have influenced creation. We are told that creation has been subjected to futility and that the fall altered creation for the worse. Yes, the heavens declare the glory of God, but we should also expect to see creation in decay.[22]

We should expect to find God's revelation in nature, not in material perfection. I am amazed, however, how often I see people adhering to the paradigm of perfection. If God made the world, shouldn't we expect it to be perfect? Like those skeptics at the crucifixion, we so easily hold opinions about God without carefully investigating his Word.

The paradigm of perfection is more a human tradition than a biblical concept. It has led to a great many unfulfilled expectations about creation. Consider again the pentadactyl pattern, which is considered evidence for evolution because it is said to reveal imperfect design. Darwin argued that the same pattern should not show up where the need is different. Surely the tasks of grasping, digging, paddling, and flying should call for different designs. Darwin doubted that all this was the product of God's design.

But why should we see this as a sign that these species were not created by God? Could not God's consistent use of this pattern be a sign that they *were* created by the same hand? After all, if the designs were all different and somehow optimized for their respective applications, then evolutionists would point to that as evidence of natural selection, just as they do with the other amazing designs in biology. If God created the species, they could argue, wouldn't we see some pattern? Instead, all we see is adaptation. Why wouldn't God leave some sign that they were created, instead of making the species appear to have evolved by natural processes?

Indeed, there are multitudes of incredible designs in nature, perfectly fitted to the need. Consider *Anableps anableps,* a fish that swims with eyes half in and half out of the water. As with bifocal lenses, its eyes are divided into two parts, giving it the remarkable ability to look simultaneously above and below the water line. Or again, there is the ancient trilobite. It had eyes that were perhaps the most complex of all.[23] One expert called them "an all-time feat of function optimization."[24]

There certainly is an abundance of perfections to be found in nature that reveal the glory of God. Literally volumes upon volumes could be written on biology's designs that can be said to be optimal or perfect. Why then are repeated designs, such as the pentadactyl pattern, such strong evidence for evolution? Is the vast abundance of perfections and contrivances insufficient to believe that God made the species? Do we need yet one more proof?

At the crucifixion they called for yet one more proof. Jesus was to save himself. Then, they declared, they would believe. But as Jesus had pointed out in the parable of Lazarus and the rich man, this was not true. There already was an abundance of reasons to believe. One more miracle would change nothing. Three days later Jesus would prove it.

10

Come Let Us Reason

The Intelligent Design Theory

If creation serves as general revelation, then what does it tell us about God's creation acts? The intelligent design theory, or ID, claims that creation reveals evidence for design. Indeed, the closer we look, the more obvious are the signs. Even evolutionists agree that life has the *appearance* of design. But the design does not always promote immediate and earthly happiness. And this makes for one of the favorite arguments against ID. "If the designer is so clever," the evolutionists ask, "why do we have parasites?"

This criticism reveals, yet again, how evolution relies on metaphysical premises. But this is just the beginning of the intense criticism that ID has drawn from its detractors. There is plenty more criticism, but unfortunately most of it lacks substance.

For instance, another favorite criticism is that ID does not qualify as a genuine scientific theory. No less a public figure than Senator Edward

Kennedy has made this objection.[1] Senator Kennedy did not provide any rationale for his claim. This is not surprising, since even professional philosophers who study science have a difficult time describing its boundaries. Just what is and is not science is hard to say. But there is no question that the design inference, which is at the heart of ID, is commonly used in science.

One idea is that scientific ideas must be falsifiable. The great philosopher Karl Popper developed this criterion and at one point used it to question the scientific status of evolution. But now one finds evolutionists liberally applying Popper's criterion to ID. Iain Murray, for example, claims that ID, ultimately, is not falsifiable. Whether Murray is correct is a point of contention. What is not in doubt, however, is that Murray's point is of little consequence, because Popper's criterion doesn't work very well in the first place. There are many endeavors that are clearly scientific but also not falsifiable.

Another idea is that scientific ideas generally fit into paradigms. Thomas Kuhn developed this idea, and Murray uses it as well to judge ID. The problem, Murray tells us, is that "intelligent design is not part of any current scientific paradigm, and besides a few fringe elements, no serious evolutionary biologists accept it."[2] Again, Murray's understanding of the philosophy is a bit lacking. Kuhn's idea includes the concept of paradigm shifts. They are messy affairs, and thinkers in the old paradigm do not easily accept the shift. Whether ID will become a strong paradigm is yet to be seen, but the fact that most evolutionists don't accept it is of no surprise and hardly serves as reason to reject ID at this point.

A Hidden Religious Agenda

The most popular criticism of ID is that it entails a hidden religious agenda. In fact ID is often simply equated with biblical creation. John Rennie of *Scientific American*, for example, refers to ID as a "creationist idea,"[3] and physics professor Adrian Melott calls ID "the cutting edge of creationism."[4] Professor emeritus of philosophy Andrew Oldenquist says ID "is not a competing biological theory; it is a religious doctrine,"[5] and an editorial of the *Columbus Dispatch* bluntly calls ID "the new name for that old-time creationism."[6]

The implication is that anything that entails religious ideas cannot qualify as science. As the *Dayton Daily News* put it, ID "isn't a scientific theory; it's a religious belief."[7] The dividing line is simple. Science, it is assumed, is somehow aloof from religious ideas. It is an objective, neutral activity that is unbeholden to metaphysical assumptions. Religion, on the other hand, is tainted. It is based on subjective, indefensible assumptions that have no place in science.

It is disappointing that this "science vs. religion" myth persists, especially after the good scholarship of the last half century that has shown how this view fails both historically and philosophically. Not only have scientists engaged in theological speculation since modern science began, but ID, as it has been formulated, is not particularly heroic, metaphysically speaking. The idea that complexity, organization, self-assembly, and the like can be interpreted as evidence for design is hardly an unscientific or unreasonable hypothesis.

If evolutionists are trying to stamp out excessive religious speculation within science, they ought to look closer to home. As we have seen, evolution is committed to far more extravagant religious assumptions than is ID. While design theorists are looking at the most complicated structures in the universe and wondering if they weren't designed, evolutionists are employing assumptions about how God must have created the world in order to verify their unlikely idea that these complicated structures somehow arose on their own. They have little idea how this could occur, but they're sure it wasn't by design. Now tell me where is the hidden religious agenda?

The Privatization of God

If we look carefully at the criticism of ID, we can see why it is so harsh. ID, rather than simply foisting its own religious assumptions is, instead, by its very nature, contradicting the religious assumptions of evolution. I have seen scientific arguments get heated, but they are nothing compared to religious arguments.

The motivating idea behind evolution is the paradigm of perfection. God "wills and wishes the happiness of his creatures," and he seeks to construct an optimal world. Since the world does not meet with our expectations, we imagine that God used natural laws rather than direct intervention. The natural laws, left to themselves, are second-rate. It is reasonable that they would create a world full of "errors and bungles," as Ralph Cudworth put it.

A corollary to this idea is that God is undetectable in creation. For natural laws, even if God used them as his creation tools, are still just natural laws. As such, his creative acts were done in such a way so as to have the appearance of having been done strictly by unguided natural forces. We may believe that God created the species directly, but only in an undetectable way.

This leaves science free to go about its investigations without having to consider the God hypothesis, and it has the effect of separating God from the world. After all, if God's only potential involvement with the world is otherwise undetectable, then why consider it? This separation of God and the world is one aspect of Gnosticism. It is not surprising that these ideas are encouraged by evolution. As I discussed in *Darwin's God*, Gnostic ideas predated and influenced the development of evolution, and the wide acceptance of evolution, in turn, strengthened modern Gnosticism.

Today, these ideas have had the effect of privatizing God. Evolution has helped to advance the notion that matters of faith should be kept private and out of public life. The reason is that if God is separate from the world and cannot be objectively verified, then what we believe about God is strictly subjective—a matter of opinion.

Those who promote this view claim it is neutral and fair to all, for those who wish to believe are free to do so. Likewise, those who wish not to believe are free from unsolicited exposure to religious ideas. God need not be acknowledged in public, for faith is a private affair. Indeed, God *should not* be acknowledged in public, for this inevitably would force one person's religion on another person.

In America these ideas have resonated with the secularization of the government. There is now firmly entrenched a doctrine of separation of church and state. It is commonly interpreted as the idea that the government may not support or allow any type of religious activity. And the government includes everything from the White House to the local elementary school. God has now been privatized in America.

The problem with this view is that it is not religiously neutral as claimed. It is, in fact, wedded to its Gnostic roots as firmly as ever. What is more, its advocates are not generally able to understand the religious bias that is woven into their view. They are apparently so deeply Gnostic that they cannot perceive their own religious position. To them their position seems to be religiously neutral.

Why is the privatization of God not religiously neutral? The simple answer is that it presupposes that God *can be* privatized. While its advocates think they are being neutral because they are allowing for the existence of God, they are allowing only for a God who isn't involved in the daily matters of our lives. This is the god of Gnosticism, a god who is disjointed from the world. They are rejecting Anselm's God who created and intimately governs the world; the God who created the nations and installs their leaders; the God who has a plan for our lives.

This is the God whom Solomon referred to when he wrote that the "fear of the Lord is the beginning of knowledge, but fools despise wisdom and discipline."[8] And this is the God to whom Paul referred when he wrote "in whom are hidden all the treasures of wisdom and knowledge."[9] This God is not a God that humanity scrutinizes and evaluates. Our wisdom does not *lead to* this God, it *begins with* this God.

This God is the God that humanity rejects. This God is why the Gnostic idol is so preferable to fallen humanity. The movement to privatize God is opposed to this God; it is not neutral. Far from embracing a separation of church and state, America has embraced a nonbiblical religion. And this is why the criticism of ID is so harsh. For the premise behind ID is that design is detectable in creation. Even though ID is careful not to describe a creator, it nonetheless violates Gnostic assumptions about God and the world. By claiming that design is detectable, ID violates the Gnostic premise that God must be separate from the world and not detectable, an object of pure faith.

What Intelligent Design Is and Is Not

ID is a research program that claims design is evident in biology. What exactly does this mean, and how is it different from evolution? The concept of design, per se, is not peculiar to ID. Even evolutionists admit that organisms show the appearance of having been designed. This appearance of design, they say, is due to the evolutionary process. Evolutionists sometimes use design-type language when speaking of the process of evolution.

The word *intelligent* distinguishes ID from evolution. It means that the design we observe was brought about by an intelligence rather than the interplay of unguided natural forces. ID claims that an intelligence is necessary to account for at least some of the design in biology. In other words, the evolutionary explanation is not sufficient.

Why is evolution unable to account for the design in biology? Because there are examples of design that cannot have arisen via a sequence of functional intermediates.

Evolution requires that every structure in biology, whether at the molecular or the morphological level, arose via a series of functional intermediate forms. In order for evolution to create the appearance of design, functional intermediates between all designs must exist. And the sequence of intermediates must be sufficiently gradual so that each successive form could have arisen from the previous form via normal biological variation. ID claims that at least in some cases no such sequence is possible.

There are several common misconceptions about ID that need to be cleared up. First, there is the notion that with ID, biological designs must be optimal in some way. ID makes no statement about optimality of designs. Such a question would mean defining the optimality criteria—in what way are the designs optimal? Obviously there are many different ways to judge a design, and ID does not address these. In contrast to the paradigm of perfection, ID merely claims that there are biological designs that require an intelligence and nothing more. Indeed, a deadly parasite could conceivably be evidence for ID.

The second common misconception is that ID does not allow for any kind or degree of evolution. ID does not reject evolutionary change altogether; it merely rejects the idea that such evolutionary change could do the job all alone. We might say ID rejects evolution but not evolutionary processes. In fact, as we shall see, there are interpretations of ID that allow for quite a bit of evolutionary change.

This is in contrast to evolution, which does not allow for any intelligent design. This is because, as we have seen, evolution is based on religious premises that ID violates. If design is detectable—if there is a structure without functional intermediates—then evolution as a comprehensive theory or fact is false, though much evolutionary change is still possible. According to ID, evolution cannot explain all designs, but evolutionary processes are not ruled out. According to evolution, ID must be ruled out altogether.

The third common misconception is that ID does not make scientific predictions. ID, it is said, does not encourage further research because it simply accepts what is observed as the act of a higher intelligence. Instead of searching for the reasons why, ID merely observes. This is simply a mischaracterization of ID. And coming from evolutionists it is ironic, given evolution's battery of explanatory devices that are used to reconcile the theory with what we observe in nature. Even the DNA code can be explained in terms of evolution.

Furthermore, evolutionists have no such monopoly on intellectual and scientific curiosity. Contrary to their claims, they are not the only ones who are able to research and investigate nature's inner workings. The history of science is full of researchers who believed they were investigating a creator's design but nonetheless were not somehow stifled from further work.

Design-based Research

The idea that the design perspective cannot stimulate research or formulate predictions reveals a misunderstanding of the perspective. The technical journals are full of research that is based on the design premise. The design perspective opens up a wide range of research areas and predictions in the life sciences. One such area is design topology. For example, very different amino acid sequences can make for the same protein. Hemoglobin, for instance, can be produced from sequences that have practically no more similarity than would be expected from two random sequences. And within the family of all globin proteins, there is a great variety of sequences. These sequences form clusters. What sort of region is defined by the set of all globin sequences? Is there a large, single region in sequence space that contains all these sequences and more? Alternatively, are these clusters connected by narrow bridges such that the known clusters constitute the majority of the region? Or do the clusters form isolated islands? Recent experiments and analysis argue against the large region model, but there is much more to learn, and the results will be relevant to protein design and the biotechnology industry.

A related area of research involves the question of why those different sequences are used. Evolutionists typically view them as the result of random changes. In other words, there is no functional reason for the differences. This is typical for evolutionary theory. Rather than search for a function, evolutionary theory quickly concludes that a design is vestigial or perhaps the result of neutral evolution. In this way evolutionary theory, not ID, stifles research. The result is an "evolution of the gaps" theory. Gaps in our knowledge are explained as the result of evolution. If we have no knowledge of a function, then the unguided process of evolution created it. But this explanation is steadily pushed into the corner as our knowledge increases and we continue to find new functions. As we saw in chapter 4, for example, Robert Wiedersheim claimed in 1895 that eighty-six organs in the human body were vestigial, but twentieth-century biomedical research has found functions for practically all of them.

ID will encourage the search for function at the morphological as well as the molecular level. In our globin example, the question is why all those different sequences are used. This is not an easy question, for it will require an understanding of the workings of the entire organism, or at least an entire cell. At the molecular level it appears that very different sequences could be freely substituted, but at the cellular level it could be that the sequence differences are there for a reason. This research will help us understand the molecular-to-morphological connection, a key to a deeper understanding of biology.

Designs that are repeated in otherwise different species are not a problem for design-based research. There are plenty of such examples in biology, and evolution must liberally make use of convergent evolution as an explanatory device. An outstanding example is the marsupial-placental convergence in mammals that we saw in chapter 4. Evolution must explain this unlikely set of duplicated designs as evolution repeating itself. On different continents and over millions of years, the blind forces of evolution are supposed to have found practically identical solutions over and over. These repeated designs are naturally explained by ID, without having to resort to just-so stories. At last we will have a framework that allows researchers to explore function and design without having to force-fit results into an unlikely scheme.

Similarly, design-based research can readily accommodate small-scale evolution. Instead of having to imagine that small-scale change must somehow extrapolate to massive amounts of large-scale change (in spite of the empirical evidence), we will now be able to see it for what it is. Small-scale change can be viewed as a mechanism for adaptation and preservation. And it also can be seen as a great opportunity. Scientists can research and implement small-scale change for the good of humanity. Higher crop yields, freeze- and pest-resistant crop varieties, ecological control and habitat recovery, vaccines and healthier livestock are just a few examples of what can result when our research focuses on the productive uses of small-scale change.

Indeed, much of today's life science research work focuses on design and function. Because Darwin's theory of evolution is dominant, the work is cast into this paradigm, but this is not the natural paradigm. Much of our current life science work fits better into the design paradigm. Though ID does not reject the evolutionary process, there are substantial differences between the two paradigms. Where evolution will accept and even look for nonfunction, ID will look for function. Where evolution will explain away the obvious designs in nature as chance products of natural selection, ID will simply model the design as design.

Metaphysical Interpretations of Intelligent Design

Quantum mechanics raises certain metaphysical questions, and there are different approaches to answering them. Likewise, ID raises certain ultimate questions related to the origin of species. Just as quantum mechanics can be taught without delving into the metaphysics, so too can ID. But the questions remain and must be considered by philosophers and theologians. ID is not a "theory of everything." It does not dictate the ultimate truth, but rather allows for a broad spectrum of diversity. The metaphysics behind ID can range from *ex nihilo* creation to *design via secondary causes* and everything in between. A detailed treatment of

123

these ideas is beyond the scope of this book. Here I will briefly summarize four very different approaches.

Ex nihilo and *de novo* creation mean the creator formed the species out of nothing, or out of nonliving matter, respectively. The species were not formed as derivatives from preexisting species. This is the most interventionist of all approaches, and there has been substantial religious feeling against it for this reason. Many feel that God would not be so involved with the details of creation. The fact that the world is not always harmonious has served to increase the opposition. Despite its metaphysical opposition, I believe this approach continues to provide the best empirically-based and parsimonious explanation for the origin of species.

An approach that requires slightly less intervention might be called *descent with design*. Here the evolutionary process is modified or guided along the way with exterior inputs. Design is injected into the process. This idea is motivated, at least in part, by the paradigm of perfection. What we believe are suboptimal designs are viewed as designs that have not yet been updated or replaced. For example, similarities in different species that do not seem optimal are viewed as unmodified by the design process.

An approach with even less intervention is the *front-loaded* creation idea. Here, all the design is injected into the first living cell (or cells), and the evolutionary process takes over from there. The potential for all the species is implicit in the first organism, and it is realized by the action of natural laws.

Finally, there is *design via secondary causes*. Here there is no detectable injection of design. Design is not imputed all at the beginning or at discrete points along the way. Instead, the design is in the initial arrangement of matter and the action of natural laws. And we should not underestimate the power of those natural laws, given quantum mechanics, chaos theory, and who knows what else that will be discovered in the future. Those laws may be able to control and manipulate creation in far more subtle ways than we have imagined.

Indeed, some may argue that *design via secondary causes* is an interventionist approach every bit as much as *ex nihilo* creation is. For instance, weather systems continually evolve. On the one hand, we say they move according to natural laws, but because they are so chaotic we cannot say a divine hand is not controlling them. Natural laws and the systems they operate on are so complex that it could be that God can actively control the world without violating what we perceive to be the actions of natural laws.

These are but a sampling of the metaphysical ideas that lie behind ID. Each idea can be said to be consistent with biblical creation, although different levels of symbolism may be required. For example, Genesis tells us that Adam was made before Eve. This poses no problem for *ex nihilo* creation but must be read allegorically in the other approaches.

Because Darwin's theory of evolution is currently dominant in the life sciences, it may seem natural to assume that *ex nihilo* creation is less scientific and more religiously motivated than the other approaches. But as we have seen, evolution has its own religious motivations and is full of scientific problems. Indeed, though the evaluation of scientific evidence is ultimately subjective, I believe it supports *ex nihilo* creation better than the other approaches. It is, I believe, the religious motivations rather than scientific motivations that have kept those other approaches popular.

In any case, all of these approaches require God's miraculous works. Yes, Darwin's sentiment was true: there is a grandeur to life. But Darwin's focus on natural laws, to the exclusion of all else, has failed us. Creation is a complex, intricate web that cannot be reduced to the product of a set of natural laws. With the grandeur there is a marvelous diversity. Yes, we can model much of what we see as the workings of natural forces, but there must also be a creator who designed and uses the laws. The world is not a chance event. It is a divine creation. This postulation is no more religious and is even more plausible than evolutionary dogma. Anselm's proof, recast and reconsidered for the twenty-first century, overtakes Darwin's proof. Our lives—and our science—can be the richer for it.

11

Wisdom Rejoices:
The Nature of Design

In telecommunications theory the word information is equated with unpredictability. This is because a word or a character in a message that is predictable doesn't inform. It carries less information than a character that is unpredictable. Information is maximized when a message is full of unpredictable characters. This approach has helped us to learn how best to store and transmit data, but telecommunications theory has its limits. For example, it would judge a meaningless message of randomly generated characters to have maximum information content. Likewise, it would judge messages we find important—everything from an automobile repair manual to a Shakespeare play—to be low on information because they are, to a certain extent, predictable and nonrandom.

Meaningful messages must be somewhat predictable. This is because languages and meaningful messages that use them must have certain patterns that make for predictability. Languages are not systems where all outcomes are equally probable. To be understandable, to convey meaning, a written message must conform to rules of spelling, structure, and sense. Also, meaningful concepts require a context. Ideas must be introduced, described, compared, distinguished, and so forth to convey a message. Thus the message will repeat certain words and phrases in building the overarching idea.

A good mystery novel will surprise the reader but it must not be so bizarre that it is implausible. Short-story author O. Henry (William Sydney Porter) was famous for his surprise endings, but even his plots fell within certain constraints. A story consists of characters, relationships, and a plot line. It constructs a context within which characters play out the drama. However loosely constructed, it will at least have some structure.

Machine Design

A writer has great freedom but must work within constraints. The construction of stories is analogous to the construction of machines. Machines are not randomly designed. There are certain design trends that result from the fundamental operations of machines. They use energy to perform a function according to the laws of physics, so their design is constrained. Consider the set of all aircraft. There is only a small set of fuels with sufficient energy density, storability, portability, and safety for use in aircraft. These are, of course, high-octane gasolines. All aircraft must operate within the same atmosphere. The various properties of the atmosphere—its density, oxygen content, variation with altitude, turbulence, and steady winds—are parameters within which aircraft designers must work. An engine designer may want to have a more oxygen and a wing designer may want to have less turbulence, but they must work with reality.

Consider two fundamental aspects of aircraft design: propulsion and aerodynamics. Obviously both are highly constrained by the reality within which they must function. Aircraft propulsion provides the thrust force and is usually based on either piston or turbine engines. The aircraft aerodynamics provides the lift force and is more streamlined for higher speed aircraft. Not only are these two aspects of aircraft design highly constrained to certain categories of designs, they are also correlated with each other. For example, military jet fighters are designed with gas turbine propulsion and supersonic aerodynamics, whereas general aviation aircraft are usually designed with piston engines and low-speed aerodynamics.

The aerodynamics and propulsion design components are correlated. An aircraft with supersonic aerodynamics will be powered by turbine engines; an aircraft powered by a piston engine will have subsonic aerodynamics. Likewise even with seemingly unrelated design components. The actuators that control the aerodynamic surfaces, the landing gear, the window design, and the avionics are all constrained and correlated. This is well illustrated by the early and often humorous attempts at flight. A great variety of designs were tested. Some, for example, used flapping wings since, after all, it worked for birds. But most designs simply did not work. A great deal of experimenting with all sorts of design ideas was needed before people like the Wright brothers and Glenn Curtiss converged on successful designs.

It is no surprise that machine designs tend to fall into very definite patterns and categories. Only particular designs work, for the reality within which the machine must operate guides and constrains the design. If machines were messages they would contain little information, for they are highly predictable.

Despite these constraints there is nonetheless a great freedom in the machine design process. As with stories, machines can be adjusted by slight changes or by component swapping. That is, just as a sentence can be modified by altering one word, a machine design can be altered by changing one dimension. And just as phrases can be mixed and matched, for example, to make a metaphor, machine components can be swapped. Gas turbine engines can be used with supersonic or subsonic aerodynamics. Or again, airplanes usually use wheels for their landing gear, but they can be outfitted with boat-like pontoons instead, enabling them to land on water rather than on land. Yes, design components are correlated, but there are exceptions.

Design in Biology

Machine design principles also apply in biology. As with man-made machines, biological machines expend energy to function in the environment in which they find themselves. And as with man-made machines, we find an abundance of patterns and categories. Over and over nature reveals similar designs. From the molecular level to the morphologies of the species, nature presents variations on the same design. Variations of hemoglobin and myoglobin, the oxygen-carrying proteins, are found in everything from plants to annelids, arthropods, mollusks, and vertebrates. Likewise, the same forms of molecular machinery that store, retrieve, and interpret genetic information are designed into all organisms. Multicellular organisms typically use similar raw materials. Their bones and their blood, even in otherwise distant species, are often of the same forms. The visible components, even in species a world away, often follow the same design principles.

As with machine components, biological design components tend to be correlated. Organisms with similar bones will usually have similar brains, but not always. In fact the exceptions can be significant. Consider the whale. It has lungs, hair, and does not lay eggs but rather gives birth after a long gestation period to a young calve, which is nourished with milk from its mother. For these and other reasons the whale is classified as a mammal. But unlike most mammals, whales act and look like fish. They live their lives in the ocean and indeed they thrive there. Their design and remarkable underwater abilities make them prolific mariners. They feature a remarkable and effective combination of components.

As we saw in chapter 4, the marsupials and placentals are another striking example of this sort of organic word play. These two types of mammals have different reproductive systems, but otherwise we find a common set of subsystems combined in the same way over and over. It is like two car lots with similar lineups of automobiles except that the one is full of gasoline-powered cars while the other is full of diesel-powered cars. Biological variation is observed in the form of slight changes and in the form of component swapping.

Of course the mixing and matching of organic components has its limits. The organism must be viable and functional, and the realities of the organism's environment must not be ignored. The mechanism of mixing and matching is interesting because it is such an obvious indicator of design. It is precisely this sort of component swapping that one would expect of designed objects, not evolved objects.

Evolution's Contingencies

Why is the mixing and matching of organic subsystems inconsistent with evolution? The reason is that any event in the evolutionary narrative is contingent upon the immediately preceding events. The narrative thread cannot take leaps—it is limited to the action of natural forces on the biological structures that are immediately available. It cannot recruit designs that are half a world away or from another era, for it acts on the here and now.

Furthermore, at any moment in time the thread may move in one of many directions. Rarely in biology is there only one solution to a given problem. For example, Darwin, and evolutionists ever since, have pointed to repeated designs such as the pentadactyl pattern (five digits or fingers at the end of the limb) as evidence of the lack of design. After all, are five bones the best design for grasping, walking, flying, and climbing? The pattern appears in limbs that perform all these varied functions, but surely five is not the best number for all these different applications. Darwinists argue that the pentadactyl pattern is used in a variety of species because it happened to be there, not because it *had* to be there.

For evolutionists it seems obvious that there are many different solutions in biology. Things probably could have been different. Francis Crick summed this up with his characterization of the DNA code as a "frozen accident." Why was one code chosen over the others? It wasn't; it was an accident—a result of a host of factors and contingencies that we'll never know.

It seems likely that the bat could be designed to function just fine if it had a six-bone pattern rather than a five-bone pattern in its wing structure. So for evolutionists, natural history is one of a million possibilities and one reality. There is no end goal or reason why one is chosen over the others. It is "chosen" only because it happened to occur, and it occurred only because it was produced by a multitude of unguided, random contingencies. "Replay the tape of history," evolutionists like to say, "and things would be different."

The early science fiction writer Ray Bradbury illustrated this view of nature in his short story "A Sound of Thunder." A cowardly but curious safari hunter named Eckels is the main character. It is the middle of the twenty-first century when time travel has not only become possible, it is rather mundane. Eckels signs up to go on a safari back in time but doesn't respect the hazards of time travel.

The safari company has carefully selected dinosaur targets for the hunt. The dinosaur must be one that is about to die from some natural cause such as drowning in a tar pit. They may kill the beast a moment before its death, but otherwise they must leave the scene untouched. They use an antigravity metal path to walk about without touching the ground, and they even recover the bullets after the kill.

But these precautions are lost on Eckels. Travis, the safari guide, must explain to him the fragility of the future:

"A Time Machine is finicky business. Not knowing it, we might kill an important animal, a small bird, a roach, a flower even, thus destroying an important link in a growing species."

"That's not clear," said Eckels.

"All right," Travis continued, "say we accidentally kill one mouse here. That means all the future families of this one particular mouse are destroyed, right?"

"Right"

"And all the families of the families of the families of that one mouse! With a stamp of your foot, you annihilate first one, then a dozen, then a thousand, a million, a billion possible mice!"

"So they're dead," said Eckels. "So what?"

"So what?" Travis snorted quietly. "Well, what about the foxes that'll need those mice to survive? For want of ten mice, a fox dies. For want of ten foxes a lion starves. For want of a lion, all manner of insects, vultures, infinite billions of life forms are thrown into chaos and destruction. Eventually it all boils down to this: fifty-nine million years later, a caveman, one of a dozen on the entire world, goes hunting wild boar or saber-toothed tiger for food. But you, friend, have stepped on all the tigers in that region. By stepping on one single mouse. So the caveman starves. And the caveman, please note, is not just any expendable man, no! He is an entire future nation. From his loins would have sprung ten sons. From their loins one hundred sons, and thus onward to a civilization. Destroy this one man, and you destroy a race, a people, an entire history of life. It is comparable to slaying some of Adam's grandchildren. The stomp of your foot, on one mouse, could start an earthquake, the effects of which could shake our earth and destinies down through Time, to their very foundations. With the death of that one caveman, a billion others yet unborn are throttled in the womb. Perhaps Rome never rises on its seven hills. Perhaps Europe is forever a dark forest, and only Asia waxes healthy and teeming. Step on a mouse and you crush the Pyramids. Step on a mouse and you leave your print, like a Grand Canyon, across Eternity. Queen Elizabeth might never be born, Washington might not cross the Delaware, there might never be a United States at all. So be careful. Stay on the Path. Never step off!"[1]

But when Eckels sees the towering dinosaur up close he panics. "It could reach up and grab the moon," exclaims the overwhelmed Eckels when he first sees the beast. The immense and terrifying dinosaur is too much for him and he must retreat to the safety of the time machine. But in his fright Eckels slips off the path and onto the ground. What has he done to the future?

The safari guides are livid. One wants to leave Eckels there in the past, but of course that would risk even greater changes to the future. They take Eckels back to the twenty-first century with them only to discover a dead butterfly in the mud on Eckels's boot and along with it a different world.

Eckels killed the butterfly and altered the future. Bradbury's tale captures the view of natural history as a contingent process. Evolutionists may argue that more than a butterfly would be required to change the future, but the principle remains. Chance, says Kenneth Miller "plays an undeniable role in history. . . . The twentieth century could easily have been very different—the next century more different still."[2]

But if natural history is so pliable, if six fingers are as good as five, then how does evolution explain the mix and match variation we see in nature? The marsupial and placental lineages within the mammals are separated by great chasms in space and time. Over millions of years and in distant lands evolution is supposed to have created the same component designs. In a world of contingencies we are told that the same subsystems are created once and again on separate evolutionary lineages. Something like a mouse gave rise to these two groups, and in those two separate evolutionary histories arose cousin species with the same design components ranging from the wolf to the flying squirrel.

The Random Design Hypothesis

Evolution's explanation for what we observe in the design of the species is hardly compelling, but the similarities and differences between the species make perfect sense from a design perspective. Designed objects can vary slightly, and they tend to have correlated subsystems, though not always.

The biology literature has many examples of mismatches—groupings based on one type of feature that contradicts groupings based on another type of feature. Even one of the Darwinist's favorite examples, the DNA code, has minor variations that look more like component swapping than a result of common descent.

Miller has claimed that these slight variations "occur in regular patterns that can be traced directly back to the standard code," and therefore they "provide powerful—and unexpected—confirmation of the evolution of the code from a single common ancestor."[3] But in fact that pattern does not line up with the other design components. The DNA code variations are scattered across various types of organisms. For example, the UAR codon is observed to switch from "stop" to "Gln" in green algae, various ciliates, and some diplomonads. Likewise, the UGA codon is observed to switch from "stop" to "Trp" in other various ciliates and two firmicutes.[4]

Evolutionists must go to great lengths to explain the mismatches caused by component swapping in biology. Therefore one might think that they would not claim these comparisons as strong evidence for their theory. But they do. In fact, evolutionists claim that the similar groupings that arise from different characteristics are strong evidence for their theory.

Evolutionists agree that there can be significant differences between the groupings but, as they rightly point out, there is a pattern. The groupings reveal that the species are not randomly designed. There are design correlations that cannot be denied. In fact, evolutionists use statistics to show how incredibly unlikely the hypothesis of random design is. Given random design as the null hypothesis, we have no choice but to reject it.

Evolutionists say their theory is the only remaining alternative. How can evolutionists ignore the fact that the data are exactly what we would expect from design? The answer is they redefine design. If the species were created, they inform us, then the species would be randomly designed. According to Mark Ridley, if the species "were independently created, it would be very puzzling if they showed systematic, hierarchical similarity in functionally unrelated characteristics."[5]

We saw in chapter 6 how Darwin's arguments against Richard Owen's ideas led to a rejection of not only Owen's theory of a divinely directed evolution but also of Owen's notion of archetypes. Where Owen explained unnecessary similarities in biology as the result of a divine plan, Darwin explained them as the leftovers of the unguided process of evolution. In fact, Darwin rebuked Owen's idea. Darwin claimed that such patterns would, in fact, not be found in created species. The implication was that for Darwin such designs should be unrelated in different species—there should be no discernable pattern.

The contrast could not be sharper. Naturalists went from viewing these similarities as the result of design to an argument against design. Today, Darwin's notion is made explicit in the random design hypothesis. If God made the species, evolutionist's such as Ridley claim, then their designs should be unrelated except where function mandates a similar design. Biology is so complicated that species appear to have been designed but arguments like these mandate that the species were not designed. The underlying reasoning, however, is religious, not scientific.

The Triumph of Religion over Science

Evolutionists have always admitted that organisms appear to have been designed. This is true in more ways than one. Organisms are complicated, so complicated we do not understand many of their details. From conception to death the life path of an organism is full of complexities that we are still only puzzling over. What is clear is that the machine we call life did not fall into place on its own. The story of evolution requires great faith—greater than that required by the story of design.

133

Even the mechanisms that produce biological variation defy evolution. Biological variation is needed by evolution, but evolution does not explain how it produced those mechanisms that create the variation. The incredible complexity of these mechanisms was yet to be discovered in Darwin's day, but now we know better. Yes, organisms appear to have been designed.

Also, the species are related to each other in ways we would least expect if they evolved. We find striking differences in otherwise closely related species, and striking similarities in otherwise distantly related species. And the fossils reveal new forms abruptly appearing in the geological strata. Over and over the evolutionist must say the problem lies with the data not with the theory. The fossils do not form a continuous spectrum but fall into clusters, and most of the fossil species belong to a small number of distinct major groups.

Each of the scientific evidences cited in favor of evolution turn against Darwin's theory. Only superficially does his explanation unify all of biology as is often claimed. When we examine the data carefully we see that evolution forces them into an awkward fit. Darwin's theory becomes increasingly complicated and convoluted as our knowledge increases.

But the underlying motivation for Darwinism is far stronger than scientific technicalities. Problems with the paleontology or genetics are no match for the strong religious arguments that inspire and buttress evolution. For Darwinists, these arguments made Darwin's theory a fact regardless of the scientific quandaries. In Darwinism, religion triumphed over science to the detriment of both. We need to recognize and remedy this situation. Let us now fix our religion and our science.

Appendix

Faulty Arguments
for and against Evolution

The creation/evolution debate has been ongoing for more than a century. There has been plenty of time for the development of a wide range of arguments. But emotions run high, and a great many of the arguments that repeatedly appear add more confusion than anything else to the fray. I have tabulated and analyzed dozens of such arguments, and in this chapter I briefly discuss the problems with each one. There is a total of twenty-five arguments in six different categories. The categories are: "evolution is a fact," "evolution is good science," "evolution's religion is OK," "evolution is not religious," "arguments against divine creation," and "weak arguments against evolution."

Evolution Is a Fact

Evolutionists hold strong convictions about their theory. The three main categories of evidence (small-scale evolution, comparative anatomy, and the fossil record) are all seen as strong evidence. And as we have seen in chapter 6, it is typical for evolutionists to make bold claims about how the evidence mandates evolution.

So it is not surprising that Darwinists commonly refer to evolution as a scientific fact. In his biology textbook Neil Campbell writes that the "term *theory* is no longer appropriate except when referring to the various models that attempt to explain how life evolves . . . it is important to understand that the current questions about how life evolves in no way implies any disagreement over the fact of evolution."[1]

Douglas Futuyma informs us that evolution "is a fact, as fully as the fact of the earth's revolution about the sun,"[2] and Richard Lewontin says it is time "to state clearly that evolution is a fact."[3] Niles Eldredge claims that evolution "is a fact as much as the idea that the earth is shaped like a ball."[4]

The National Academy of Sciences explains that in science the word *fact* can be used "to mean something that has been tested or observed so many times that there is no longer a compelling reason to keep testing or looking for examples. The occurrence of evolution in this sense is a fact. Scientists no longer question whether descent with modification occurred because the evidence supporting the idea is so strong."[5]

To skeptics the idea that evolution is a fact seems absurd and hardly defensible. But evolutionists do defend this claim. And they have crafted their argument in such as way as to make the claim undeniable. There is no way to falsify the claim that evolution is a fact. This makes the claim less meaningful, and for the skeptic it once again reveals evolution's reliance on metaphysical claims. Let's have a look at the specific arguments in support of the premise that evolution should be considered to be a fact.

Evolution Is a Scientific Fact

If you tell Darwinists you are skeptical that evolution is a fact, sooner or later you'll hear the following argument: evolution, like any scientific fact, the Darwinist will explain, is not absolutely certain. Your understanding of the scientific method will be cast into doubt as you are politely told that science doesn't deal with absolute certainty. You should not doubt that evolution is a scientific fact merely because it does not qualify as a logical truth or because it is not certain beyond all shadow of a doubt. The theory of evolution is as firmly established as anything in science, but there is no place in science for absolute certainty.

Now it would seem that such a misunderstanding would be simple enough to clear up. There is a mountain of scientific evidence against evolution, and the evidence counted in its favor is often ambiguous. We are nowhere close to quibbling about degrees of certainty. Evolution is just plain unlikely.

Imagine an astrologer responding to skepticism with this sort of defense. "Look, nothing is certain," the stargazer would explain, "but don't you see this is a fact?" This would make no sense. Likewise, evolution does not merely fail to achieve the status of absolute certainty; it isn't anywhere close.

But too often I have found this misunderstanding persists. No sooner have you agreed that a fact need not be absolutely certain than you are asked to explain how evolution fails to be a fact. If a fact need not be completely certain, then why isn't evolution a fact?

At this point you realize the Catch-22 you have entered into. You have agreed that something can be a fact without being certain, so if evolution is uncertain, it can still qualify as a fact. Such a discussion quickly becomes fruitless. You realize that there is little chance of disabusing the Darwinist of the notion that evolution is a fact. Your only hope is to exit this illogical wordplay and appeal to the scientific evidence from which evolution is supposedly derived.

Evolution Is a Derived Fact

Darwinists like to say that evolution is no longer in question these days. There is a preponderance of evidence so great it cannot be denied. That the earth is round was an accepted fact long before space travel allowed us to observe it directly. Likewise, we do not require evolution to be directly observed to call it a fact. To think otherwise is to misunderstand the very essence of science.

Furthermore, the Darwinist will explain, and this is very important, you must understand that a scientifically derived fact is not necessarily completely understood. Gravity is no less of a fact because we haven't completely worked out all of its details. One needn't know all the details to know something is a fact. Thus evolution is both a fact and a theory—a theory to be worked out to explain the fact that is known.

The skeptic can hardly disagree with this seemingly sound reasoning, but he will soon regret it. For when he points out the many shortcomings of the scientific evidence, he will learn that these are the details that are still being worked out. The skeptic is now on an epistemological treadmill where no amount of difficulty that he points to in the evolutionary account can mean it is not a fact. These, you will be told, are why evolution is also a theory. Do you not accept that life has increased in complexity and that new species arise? Well, then, you accept the fact of evolution. And if you point out that we don't know *how* the species arose, then you will be told this has to do with the theory of evolution, and the treadmill continues.

The Darwinist will also point to the fact of small-scale evolution. It will get you nowhere to complain that small-scale evolution has never been observed to add up to anything but just that. For the Darwinist will ask if there is a limit to such change, and if so, what it is. When you confess your ignorance on both points, the Darwinist will smile and point out that your claim that evolution is limited must be mere conjecture. In this case, the details *are* required to establish the fact.

Stephen Jay Gould wrote much on this claim that evolution is a fact. He laid out the three categories of evidence establishing the fact: small-scale evolution, comparative anatomy, and the fossil record. What you won't find is any sort of formal argument or syllogism showing just how these paltry evidences are supposed to prove Darwin's incredible conclusion. You'll never find one, because there is none. Small-scale evolution, comparative anatomy, and the fossil record simply cannot bridge the immense chasm separating Darwinism from the evidence.

Furthermore, it is the hallmark of science to consider all the evidence. What about the problems with these three categories of evidence? And what about the counterevidences we saw in chapters 2 and 3? These, you will be told, are open areas of research. The fact of evolution, reassures the Darwinist, "is as well established as anything in science—as secure as the revolution of the earth about the sun."[6]

Evolution Is a Logical Fact

Darwinists also argue that evolution is a fact due to the failure of spontaneous generation—the idea that organisms could arise spontaneously. Spontaneous generation explained, for example, why rats could always be found in garbage dumps—they spontaneously arose from the garbage somehow. Spontaneous generation was disproven by Pasteur in the nineteenth century, and with its passing, biology embraced the law of biogenesis, which stated that all life comes only from preexisting life: *omne vivum ex vivo*—all that is alive came from something living.

I first discovered this argument in the writings of Arthur W. Lindsey, a Darwinist who wrote several books in the mid-twentieth century. At the time I thought the argument was rather obscure (as well as bizarre), but I've since seen other Darwinists use it as well. It goes like this: All life comes from previous life, and we know that earth's complement of species has changed over time. Therefore, it is a scientific fact that the new species must have arisen from older species. Or as one Darwinist told me, all living forms come from previous living forms, and you shoulder a heavy burden if you assert otherwise.

I was surprised to see this argument is not as obscure as I had thought. For it would seem to rely on a rather obvious flaw, namely, extrapolating beyond the experimental conditions. Yes, Pasteur showed life comes from previous life, but life had to have a beginning. Obviously biology's dictum *omne vivum ex vivo* cannot apply for all time; even Darwinists must agree with this. Indeed, biology's dictum refers to normal reproduction, not the production of a new species. It is Darwin's hypothesis that normal reproduction, occurring over eons of time, can translate into the evolution of new species.

Nonetheless, this argument from *omne vivum ex vivo* persists, and the skeptic will make little progress convincing the Darwinist of its weakness. The Darwinist will shift the burden of proof and say you must disprove *omne vivum ex vivo*. And if you say you accept the dictum but not its unlimited application, then you will be asked to define the limit. If you don't know the limit, then how do you know it is limited? As with the limit of small-scale evolution, the details *are* required to establish the fact. The Darwinist has misunderstood biology's *omne vivum ex vivo* and has used it in a way never intended.

Evolution Is a Metaphysical Fact

Finally, evolution is a fact because divine creation must be false. The reasoning, as we saw in chapter 6, involves personal religious or metaphysical beliefs. This leaves evolution as the only option. For example, about fifty years ago the paleontologist and Roman Catholic cleric Teilhard de Chardin elevated evolution to the status of a metaphysical truth: "Is evolution a theory, a system, or a hypothesis? It is much more—it is a general postulate to which all theories, all hypotheses, all systems must henceforward bow and which they must satisfy in order to be thinkable and true. Evolution is a light which illuminates all facts, a trajectory which all lines of thought must follow—this is what evolution is."[7]

Likewise, for Stephen Jay Gould it didn't seem to matter that his three categories of scientific evidence (small-scale evolution, comparative anatomy, and the fossil record) fail to establish evolution as a fact. For Gould did not limit himself to scientific interpretations of the evidence. Whenever he presented his argument for the fact of evolution, he gave the reasons why he believed these evidences disprove divine creation.[8] Gould's reasoning is no different from the arguments Darwinists have always used. It relies on religious beliefs and as such is not open to scientific debate.

The metaphysical claims of Darwinists are often explicit and obvious. Other times they are more subtle. These subtle expressions, however, often follow the same pattern. Darwinists will often claim that evolution is the only acceptable explanation for nature and its species. To the unsuspecting student this may sound like an objective statement based on nothing more than empirical observations. Actually it is a dogmatic claim, for it says that no other explanation can be true.

It is typical for scientific theories to make predictions. If theory A is true, then x will be observed. The "if-then" statement is common in science. But if x is observed, theory A is not necessarily true. For example, there may be other explanations not yet considered. The only way for observation x to prove theory A is if *only* theory A predicts observation x. In this case the simple "if-then" statement is replaced by the "if and only if-then" statement. But it is far more difficult to prove that one's favorite theory is the only explanation for an observation.

Beware of phrases such as "only evolution" and "nothing except evolution." Tim Berra claims that the theory of evolution is "the only reasonable explanation" for the fact that virtually all organisms carry their genetic information in the DNA molecule.[9] Statements like this implicitly include the assertion that divine creation cannot explain the observation. Such claims are not open to scientific debate.

Perhaps the most famous example of this sort of claim comes from Theodosius Dobzhansky. His pronouncement that "nothing in biology makes sense except in the light of evolution" has become a favorite sound bite for Darwinists, but it reveals the religion inherent in the theory. Indeed, consider these statements he made to justify his pronouncement:

> What a senseless operation it would have been, on God's part, to fabricate a multitude of species ex nihilo and then let most of them die out!
>
> They fancy that all existing species were generated by supernatural fiat a few thousand years ago, pretty much as we find them today. But what is the sense of having as many as 2 or 3 million species living on earth?
>
> Was the Creator in a jocular mood when he made *Psilopa petrolei* for California oil fields and species of *Drosophila* to live exclusively on some body-parts of certain land crabs on only certain islands in the Caribbean?
>
> But what if there was no evolution and every one of the millions of species were created by separate fiat? However offensive the notion may be to religious feeling and to reason, the anti-evolutionists must again accuse the Creator of cheating. They must insist that He deliberately arranged things exactly as if his method of creation was evolution, intentionally to mislead sincere seekers of truth.[10]

Arguments like these are commonly used by evolutionists. They are powerful for those who share these religious feelings, but for others they make little sense.

140

Evolution Is Good Science

Argument from Authority

Darwinists rightly point out that evolution is the accepted paradigm in biology worldwide. By far the vast majority of life scientists and researchers would not consider any other explanation for the origin of species. It is extremely well accepted, and, Darwinists argue, such a strong consensus indicates evolution is the best explanation we have.

The use of this argument from authority is curious, given the ethos of independent thinking in science. In my graduate studies I was taught to question everything, even the most well-accepted theories. It is good for a theory to be well accepted, but this does not mean it cannot be contested.

Of course, science cannot always be in a state of doubt. It helps for workers to operate under a common set of assumptions in order to make progress. Thomas Kuhn viewed science as operating according to paradigms. A scientist has a large incentive for accepting the current paradigm; otherwise, it will be difficult to collaborate with other scientists, publish in prestigious journals, and secure research funding. Hence, the wide acceptance of a paradigm, such as evolution, is not as meaningful as the argument from authority assumes. Ultimately what is important is how well the theory accords with the known data.

The Scientific Journal Articles Are Not Religious

Life science libraries are full of research journals that make no religious references. If the life science research and, in particular, evolution research make no reference to religion, then, Darwinists argue, how is it that evolution is religious?

In considering this question we must first understand that scientific literature consists of different genres, ranging from the intentionally dry scientific journals to popular works intended to convey scientific results to the public. No research paper would get published in a scientific journal if it made overt religious claims. The purpose of the journals is to report on empirical results, not to discuss wider implications. For example, the content of journal articles does not generally venture outside the accepted paradigm or attempt to prove the paradigm.

The religious claims of evolution are required to establish the veracity of evolution, not to operate within the paradigm of evolution. Hence, the religious claims need not be considered when doing evolution research. The religious claims arise in the apologetic works that argue for evolution; they do not appear in the journals.

Evolution Makes Successful Predictions

Evolution makes a great many predictions that we know to be true. For example, if evolution is true, then we would expect to find similarities in different species. Evolution's successful predictions, Darwinists argue, prove it to be good science.

There are two main problems with this argument. First, the predictions are not as good as advertised. Many of the predictions are problematic. For example, evolution is a very flexible theory and can accommodate a great variety of observations. In fact, in many cases evolution can accommodate opposing outcomes. Also, evolution has difficulty explaining or predicting a great many things we observe in biology.

The second problem is that it is difficult to establish the validity of a theory using its successful predictions. There are plenty of examples of theories that can make impressive predictions but are otherwise known to be false. Ptolemy's earth-centered system was able to predict the locations of planets and even eclipses, but we know the universe does not revolve about the earth.

Evolution's Religion Is OK

Great Christians Have Embraced Evolution

One prominent Christian evolutionist has said that if God decided to use the mechanism of evolution to create human beings, who are we to say that was a bad way to do it? Indeed, the Scriptures do not describe precisely how God created the species. He could have used evolution as his creation tool.

In fact, theologians have stated that Christians ought not use Scripture to reject evolution out of hand. No less a theologian than Pope John Paul II, and Pope Pius XII before him, commented on evolution, as did Princeton's B. B. Warfield, in the Reformed camp.

Both Rome and Warfield held several points in common. While stating that evolution can be consistent with Christianity, both said evolution is not a fact and that only certain forms of evolution are acceptable. Specifically, only those forms that allow the Creator to have input and to control the evolutionary process and are otherwise free of non-Christian premises are acceptable.

Clearly, neither Rome nor Warfield would sanction the sort of religious claims Darwinists use to prove their theory. And without these claims there is little reason to accept evolution. Of course there are legitimately scientific aspects of evolutionary theory, such as population genetics and small-scale evolution, that easily fall within Rome's and Warfield's guidelines. But unfortunately, some evolutionists have interpreted those guidelines more broadly than stated in an attempt to gain church acceptance.

A "Front–Loaded" Creation

Could not God have preprogrammed the history of evolution into the first living cell? In other words, God could have divinely intervened to create the first life on earth, and then God's natural laws could have acted to evolve life into the different species. Indeed, this would require an even greater God to bring about the species by the indirect action of his laws rather than by direct creation.

This idea of God preprogramming the evolutionary process into his initial creation—a "front-loaded" creation—is quite popular, especially with those who seek to reconcile theism and evolution. But front-loading is not good science. As with evolution, the idea is theologically motivated.

The problems with front-loading are, first, the lack of strong empirical evidence and, second, its reliance on an unknown mechanism for transferring all that complexity. It is not as though the front-loading idea is motivated by what we know of natural laws or by what we observe in nature. In fact, because this idea is essentially identical to evolution, it inherits all of the latter's problems.

The strength of this idea has always been that it incorporates a noninterventionist creator. This neo-Gnostic concept of God, whatever its ultimate motivation, was strongly in place long before Darwin as an attempt to explain how the world came about without divine involvement.

We saw in chapter 7 that the seventeenth-century Anglican cleric Thomas Burnet urged this view of God, for a greater wisdom would be required. And we saw that David Hume argued that God is too transcendent for us to detect his creative activities. Better to make God strictly an object of worship.

Where the natural theologians saw evidence for design in creation, Hume warned against anthropomorphizing God. Similarly, Charles Darwin argued that while the eye may appear to be designed, we ought not think that God works as man works.

Today these notions are repeated over and over. In the face of overwhelming evidence against the evolutionary process, we are told that we just don't know enough facts yet.

The front-loading idea is bad science and bad theology. It invests a great deal of power into secondary causes in order to remove the need for God to intervene against his natural laws. God's intervention is seen as theologically undesirable, so the front-loading idea is used, despite the fact that it stretches natural processes beyond their limit, or at least our understandings of those limits. That is why we must appeal to ignorance to advance this idea.

Science Always Relies on Nonscientific Assumptions

Darwinism relies on religious assumptions, but isn't this of little consequence, given that all of modern science relies on nonscientific assumptions? For example, science assumes that nature is uniform and parsimonious, but it cannot prove these assumptions are true. And many great scientists have made explicit assumptions about God in their scientific research.

For example, consider the case of the French mathematician and astronomer Pierre-Louis Moreau de Maupertuis. In the eighteenth century Maupertuis formulated his Principle of Least Action, arguing that God's natural laws should cause an economy of motion. No energy should be wasted in the path of a light ray or motion of a billiard ball. Maupertuis's principle was later refined and shown to be an expression of Newtonian mechanics and optics. Maupertuis's principle is a good one, and we don't discard it because he justified it using religious ideas. Similarly, so the argument goes, we should not discard Darwin's ideas simply because he justified them with religion.

The problem here is that Darwinism is not analogous to the Principle of Least Action. There are serious scientific problems with Darwinism, whereas the Principle of Least Action is demonstrated at the macro level over and over. There simply is no comparison between the two. Discarding Darwinism in no way jeopardizes well-confirmed theories such as the Principle of Least Action.

Darwinists rescue their theory by arguing against creation. Religious assumptions are required because the science is weak. Indeed, evolution requires a variety of special-purpose explanatory devices. This is why Darwinism is bad science. Maupertuis's Principle of Least Action, on the other hand, suffers from none of these problems.

Furthermore, it is worth noting that whereas Maupertuis openly espoused his religious convictions, Darwinists claim that religion plays no role in their theory. Maupertuis had strong empirical support and so was free to promote his assumption about a divine optimizer. Darwinists require their religious arguments and so must hide them away when they claim the high ground of science. Indeed, Maupertuis attempted to use his work as a proof for God's existence, but Darwinists are in denial about their own arguments.

Arguing against Creationism Is Legitimate

Many evolutionists have explained to me that Darwin did not come up with his concept of how God created the species by himself. He was responding to specific religious claims such as the special creation of species. If there is a strong tradition arguing that God created each individual species, it is perfectly valid to show how puzzling the facts of biology are on this hypothesis.

The problem with this argument is that evolution goes far beyond merely pointing out problems of competing theories. It is true that Darwin did not contrive the doctrine of God that he so much depended on. It was so popular in his day that even his opponents did not complain about his religious assumptions. But this does not lessen Darwin's reliance on that doctrine.

Consider how Mark Ridley presents evidence for evolution in his college textbook: "The recurrent 'laryngeal nerve,' as it is called, is surely inefficient. It is easy to explain such an efficiency if giraffes have evolved in small stages from a fish-like ancestor; but why giraffes should have such a nerve if they originated independently . . . well, we can leave that to others to explain."[11] Ridley's point is that if the evidence is against the species originating independently, then the evidence is for evolution. He says the evidence is easy to explain with evolution, but of course there are no details. The "explanation" must overcome tremendous scientific problems that he ignores.

The message is that evolution, in one way or another, must be the answer. In this sense, evolution can be anything—anything, that is, except divine creation. This is why evolution is not tied to natural selection or any other specific mechanism. It is simply anything but creation. It is *any* naturalistic explanation of the origin of species. This is why evolutionists speak of fact and theory. It is a theory in the sense that we don't know how it occurred; it is a fact because nonnaturalistic explanations (i.e., divine creation) have been ruled out.

There Will Always Be an Opposing Religious Theory

If Darwinists are not allowed to critique opposing religious theories, then all of science is in jeopardy, since there will always be an opposing religious theory. For Galileo the opposing religious theory was Aristotelian physics, which he critiqued. Does this mean that along with Darwin's, Galileo's ideas were not scientific?

The answer, again, is that the analogy fails. It is true that both Galileo and Darwin overcame long-standing traditions that required them to argue directly against those traditions. But Galileo did not have overwhelming scientific obstacles to argue around, and he would not deny his religious assumptions. Whereas Darwinists claim no religious influence in spite of their religious claims, Galileo was open about his religious influence, though his science enjoyed strong empirical support.

Yes, there will always be opposing religious theories. What is important, however, is whether the new theory can stand on its own or not.

Evolution Is Not Religious

Evolution Is Mechanistic

Darwin's theory is based on three empirical observations: (1) Biological variation is present in all populations, (2) at least some variation is inherited by offspring, and (3) organisms produce more offspring than can survive. Doesn't this make for a completely mechanistic explanation, with no religion involved?

In this argument evolutionists point out that Darwin's solution is a scientific one, that it is mechanistic and relies only on natural laws. They seem to think this makes their theory scientific. The problem is that anyone can dream up a mechanistic solution for practically anything. That doesn't make it good science. There needs to be good, compelling evidence to support the idea. One cannot separate the theory from its evidence.

We could say the universe was created by vortices. This is mechanistic, but is it likely? The religion inherent in evolution surfaces when Darwinists present their evidence.

Religion Is Not Taught in Biology Class

Biology students need not agree with, or even be aware of, any religious claims in order to learn and understand evolution. This is because religion is not taught in biology class. How then can evolution be said to be religious?

The answer is that the religious influence in evolution can be quite subtle. When the teacher says that the universal genetic code is compelling evidence for evolution, it is not obvious that a nonscientific premise is required. The student, and even the teacher, will not likely be aware of the various problems with the evidence. Nor will they likely be aware of the underlying reasons why the observations have been handed down as compelling evidence.

146

Certainly this is not peculiar to evolution. Everything from theology to quantum mechanics can be taught at introductory levels, where important details are left out to simplify the learning. Just because a young student is unaware of evolution's metaphysical commitments doesn't mean those commitments don't exist.

Methodological Naturalism Is the Basis for All Science

Evolution, Darwinists explain, is based on methodological naturalism, which is the basis for all science, from astronomy to zoology. Methodological naturalism assumes that the natural world can be explained as the result of only natural causes. Science cannot test explanations about the supernatural and methodological naturalism is completely silent on the subject of God or other supernatural forces.

This argument is self-contradictory. If methodological naturalism assumes the natural world can be explained as the result of only natural causes, then it is *not* silent on the subject of God and supernatural forces. Under this definition, methodological naturalism assumes that God did not actively create the world in detectable ways.

Furthermore, the claim that the natural world can be explained as the result of only natural causes is *not* scientific. The idea that the natural world has no detectable supernatural causes lies outside of science. It is metaphysical and cannot be verified by science.

Arguments against Divine Creation

Creation Requires an Infinite Regress

Divine creation explains the world's complexity and evidence of design as the result of the Creator. The world is complex, so it must have been created. But doesn't this mean that the Creator is also complex, and so must have been created? Using this logic, we would then need to ascribe the intelligence of the Creator to an even greater Creator, and so on. Divine creation leads to an infinite regress of Creators.

David Hume used this argument against natural theology over two centuries ago in part 4 of his *Dialogues concerning Natural Religion,* and it remains popular today. "How therefore shall we satisfy ourselves," asked Hume, "concerning the cause of that being, whom you suppose the author of nature?" Is it not arbitrary to stop at the first Creator? "If we stop, and go no farther; why go so far? Why not stop at the material world? How can we satisfy ourselves without going on ad infinitum? And after all, what satisfaction is there in that infinite progression?"[12]

Today this argument appears in e-mail chats and letters to the editor, and it appears in evolution textbooks. For example, Mark Ridley informs the student that an "unabashedly religious version of separate creation would attribute the adaptiveness of living things to the genius of God; but even this does not actually explain the origin of the adaptation, it just pushes the problem back one stage."[13] Ridley can accept that an omnipotent, supernatural agent could create well-adapted living things. But the problem, says Ridley, is to explain adaptation, and the supernatural Creator already possesses this property: "Omnipotent beings are themselves well-designed, adaptively complex, entities. The thing we want to explain has been built into the explanation. Positing a God merely invites the question of how such a highly adaptive and well-designed thing could in its turn have come into existence."[14]

So the design argument appears doomed from the beginning. To be consistent it must not stop at the first Creator, but if it continues, there will be no end and thus no satisfying solution to the existence of complexity.

This infinite-regress objection is revealing. Yes, there is a dilemma, but it is not with divine creation. In making this objection the evolutionist has already rejected the possibility of the God of the Bible, who is eternal and not created. When the evolutionist assumes that "omnipotent beings are themselves well-designed," the dilemma arises. The evolutionist is faced with a choice between two absurdities: either design and complexity arise on their own or there is an infinite regress of designers.

Evolutionists argue for the first absurdity and attempt to ascribe the second absurdity to the idea of divine creation. They employ a false dilemma to argue for their theory, but this merely reveals the problem with their own presuppositions.

What about Creation's Religious Assumptions?

The existence of a Creator is indispensable to the doctrine of divine creation. Indeed, the Creator is taken as a given, while such is not the case for evolution. Therefore, divine creation, rather than evolution, lies outside of science.

When Darwinists make this argument they fail to understand both their own theory and divine creation. All of the strong arguments for evolution rely on assumptions about God. Evolution incorporates and relies on a nonbiblical concept of God. It is simply not true that evolution is free of God.

The idea that the world was created, on the other hand, is certainly a reasonable inference. No one denies that the complexity in the world suggests design. Therefore, it seems that the idea of divine creation does not presuppose God, or at least the qualities of God, in the way that evolution does. The argument that God is presupposed by divine creation while such is not the case for evolution would fare better if inverted. It would be more accurate to say that evolution, rather than divine creation, relies on assumptions about God.

Creation Is Not Science Because It Fails to Provide a Naturalistic Explanation

Creation is not science because it does not provide a mechanistic, repeatable explanation that can be tested. It is fine for people to believe in creation, but it does not belong in science.

Aside from the fact that this claim is not supported by the history of science, the more important problem is that it is ultimately circular. What if God made the species? To this the evolutionist will respond that if we find evidence for God then we will change the way we do science.

But we have plenty of evidence for God, far more than for evolution. Yet nothing has changed, and the reason is that the evolutionist assumes up front that science must be mechanistic. Failures of evolution are always viewed as research problems—the right naturalistic explanations has not yet been found. The obvious evidence for God is forced into the awkward theory of evolution. If we restrict ourselves to only naturalistic explanations, then that is what we will find, no matter what the evidence.

God of the Gaps

Evolution, Darwinists explain, analyzes nature without resorting to miracles. There is plenty of research to do, as many uncertainties remain. But we should not simply fill in the gaps in our present knowledge with the explanation that God did it. This will damage both science and faith. On the one hand, we will stop asking the questions that lead to important scientific discoveries. On the other hand, God becomes a mere placeholder that will be increasingly displaced as we gain more knowledge. If Copernicus had been content to believe that God placed the earth in the center of the universe, then he never would have discovered that the sun sits at the center of the solar system. Let's not make our God a God of the gaps.

The hidden premise in this argument is that evolution has more or less successfully described the origin of the species. According to this premise, more research is needed to fill in some of the details, but evolution has been sufficiently confirmed such that there is no need to believe that God must have created the species. It was a mistake to use God to account for gaps in our knowledge, and we should not continue to repeat this mistake.

In short, this premise assumes that evolution is a good scientific theory for the origin of species. As such, this argument begs the question. It presupposes evolution and arrives at the conclusion that we should not use God to explain the origin of species. Of course it is not a good idea to think miraculous divine intervention occurred where natural processes will otherwise do the job. But in this case evolution doesn't do the job, and that is the problem with this argument.

The problem with evolution is not that it is slowly eroding grounds for faith in the Scriptures. Evolution makes nonscriptural assumptions from the outset—it does not discover difficulties with Scripture through some neutral investigation. The problem, rather, is that evolution is not supported by the science.

Weak Arguments against Evolution

There Is No Evidence for Evolution

Skeptics sometimes claim evolution lacks evidence. But in fact there is plenty of evidence for evolution. The evidence for evolution has many weaknesses, and there is plenty of evidence against evolution. But this does not mean there is no evidence for evolution.

It is important to understand that the existence of evidence, in itself, means little. There is an abundance of evidence for all sorts of discredited ideas. What is important is not the implications of the positive evidence, but the implications of all the evidence taken together. There is plenty of evidence the earth is flat, but the sum of the evidence tells us otherwise.

Evolution Is Impossible

Ever since Darwin, skeptics have, from time to time, attempted to prove that evolution is impossible. What better way to dispose of Darwinism? But Darwin's theory is quite flexible, as its only real requirement is to stick with naturalistic explanations. Beyond that, Darwinists are free to appeal to any explanatory device. Therefore, Darwinists can find a speculative explanation for just about any problem with the theory. The skeptic who tries to prove evolution is impossible will be frustrated, because the skeptic must demonstrate that no explanation is possible for a theory that sets practically no bounds for itself.

Beyond the fact that evolution is so undefined as to be essentially unfalsifiable, there is another reason why the skeptic should avoid this argument. Evolution claims to be a scientific theory and as such should be reasonably probable. Scientists never claim their theory is good simply because it has not been proven wrong. This would open science to all sorts of imaginative and contrived explanations for observed phenomena. Science has a much higher standard for its theories than merely not having been proven wrong. Therefore, the skeptic is assuming an unreasonable burden when attempting to prove evolution to be impossible.

Evolution Is Not Falsifiable

Given the difficulty in proving that evolution is impossible, it would seem that evolution is not falsifiable. No less a philosopher of science than Sir Karl Popper made this claim in the 1960s.

I do believe that evolution is, for all practical purposes, not falsifiable. But this argument is to be avoided because it leads to complicated and obsure philosophical discussions. Theory falsification is not a simple concept, and after stating that evolution is not falsifiable, Popper apparently retracted his claim.

Furthermore, this argument draws attention away from the very powerful arguments against evolution, such as its lack of scientific support and reliance on religious assumptions.

One philosophical problem is that of determining what constitutes legitimate falsification criteria. For example, one could say that the law of gravity would be falsified if an apple shot up into the sky instead of falling to the ground. While this indeed would falsify our understanding of gravity, it is not a very good test, because it is unlikely to occur. Nonetheless, evolutionists use such unlikely tests to argue that their theory is falsifiable. For example, if the skeleton of a human were found in the most ancient geological strata, then evolution, they say, would be proved wrong. But we know such a find is highly unlikely.

Another problem with trying to make the argument that evolution is not falsifiable is that some very respectable philosophers have pointed out that scientific theories need not be falsifiable in the first place. If you have a multitude of empirical observations supporting a theory, then it shouldn't be dropped in light of a few contradictions. Therefore, even if one were to show that evolution is not falsifiable, it would not necessarily damage the theory.

151

Evolution Is Antireligious or Atheistic

Darwin's theory of evolution has had a profound impact on society. Its claim that God is not needed to explain the origin of species has influenced many, and it certainly seems reasonable to say that evolution is antireligious or atheistic. For example, evolution always opts for naturalistic explanations, no matter how unlikely, rather than admit any possibility of God. Does this not mean evolution is antireligious?

The problem with this argument is that it misses the historical roots and underlying motivation of Darwinism. The motivation behind Darwinism is *religious*, not *anti*religious, and this makes a tremendous difference in how one understands the theory. Darwinism is the product of a long tradition of religious doctrine. Though not biblical, this doctrine has always been popular in the church. This non-Christian thought can be found in many influential figures leading up to Darwin, and it remains popular today. It involves a nonbiblical version of God, who is distanced from the world. The divine attributes of wisdom and goodness are emphasized over those of providence, immanence, and judgment.

This explains why the idea of the world arising on its own was popular long before Darwin codified its place in biology. And this explains why evolution is popular with many in the church today. If nothing else, the fact that believers accept evolution shows that evolution is not necessarily atheistic.

Evolution Provides No Basis for Truth

If evolution is true and humanity evolved from an unguided process, then the human brain is nothing more than a sophisticated computer. We take in data through the senses, and the brain produces electrochemical outputs. There is nothing to guarantee the accuracy of these signals. Indeed, with evolution there is no basis for truth, as the universe is nothing but matter and energy. So isn't evolution self-refuting? Darwinists claim that evolution is true, but how could they know this? If the theory is true, then there is no basis for truth—if it is true, we would never know such a thing.

The problem with this argument is that it equates evolution with atheism and materialism. As discussed above, this is not an accurate portrayal of Darwin's theory. Though many Darwinists have used evolution to argue for materialism, many others can agree that God exists, though he must be remote and uninvolved. Thus, Darwinists can borrow whatever metaphysical foundation they require from theism without sacrificing their theory. Again, in order to understand Darwinism and make cogent arguments against it, we must understand its religious foundation.

Evolution Is Just a Scientific Theory

Evolutionists claim that evolution is a fact, but some skeptics say it is just a theory. In fact, they often go out of their way to point out that evolution is a scientific theory. Their point is that evolution is not a fact, but merely a theory that could turn out to be wrong.

The first problem here is that some scientific theories are so well accepted that it is reasonable to consider them to be a facts, though the details are not all well understood. The second and more important problem here is that this argument misses the bigger picture. It is the wrong question to be asking. We need to ask whether or not evolution is scientific at it core. Once we understand that evolution is ultimately a religious theory, then it becomes clear that evolutionists claim it to be a fact because they are absolutely certain of it in the religious sense.

Notes

Chapter 1

1. Ernst Mayr, *What Evolution Is* (New York: Basic Books, 2001), 264.

Chapter 2

1. Benjamin Lewin, *Genes VII* (Oxford: Oxford University Press, 2000), 292.
2. Ibid., 294.
3. Bruce Alberts, Dennis Bray, Julian Lewis, Martin Raff, Keith Roberts, and James D. Watson, *Molecular Biology of the Cell*, 3d ed. (New York: Garland Publishing, 1994), 241.

Chapter 3

1. Enrique Meléndez-Hevia, Thomas G. Waddell, Marta Cascante, "The Puzzle of the Krebs Citric Acid Cycle," *Journal of Molecular Evolution* 43 (1996): 293–303.

Chapter 4

1. George G. Simpson, *Horses* (Oxford: Oxford University Press, 1951), quoted in Richard Milton, *Shattering the Myths of Darwinism* (Rochester, Vt.: Park Street Press, 1992), 102.
2. Tim M. Berra, *Evolution and the Myth of Creationism* (Palo Alto: Stanford University Press, 1990), 31.
3. Ibid., 50.
4. Kenneth R. Miller, *Finding Darwin's God* (New York: Cliff Street Books, 1999), 43.
5. Quoted in Charles C. Gillispie, *Genesis and Geology* (Cambridge: Harvard University Press, 1951), 135.
6. Niles Eldredge, "An Extravagance of Species," *Natural History* (American Museum of Natural History) 89, no. 7 (1980): 50.
7. T. S. Kemp, *Fossils and Evolution* (Oxford: Oxford University Press, 1999), 16.

155

8. Robert Carroll, *Patterns and Processes of Vertebrate Evolution* (Cambridge: Cambridge University Press, 1997), 8.

9. Ibid., 9.

10. Ibid.

11. Ibid.

12. Ibid.

13. Ibid., 10.

14. Niles Eldredge and Stephen J. Gould, "Punctuated Equilibria: An Alternative to Phyletic Gradualism," in *Models in Paleobiology*, ed. Thomas J. M. Schopf (San Francisco: Freeman, Cooper, 1972).

15. Henry Gee, *Deep Time: Cladistics, The Revolution in Evolution* (London: Fourth Estate, 2000), 1–2 (emphasis in original).

16. S. R. Scadding, "Do Vestigial Organs Provide Evidence for Evolution?" *Evolutionary Theory* 5 (1981): 173–76.

17. The Wistar Institute, "Essential Cell Division 'Zipper' Anchors To So-Called Junk DNA," *Science Daily*, 30 August 2002, http://www.sciencedaily.com/releases/2002/08/020830072103.htm.

18. Gavin De Beer, *Atlas of Evolution* (London: Nelson, 1964), 38.

19. Edward O. Dodson and Peter Dodson, *Evolution: Process and Product* (New York: D. Van Nostrand, 1976), 51.

20. Berra, *Evolution and the Myth of Creationism*, 22.

21. Mark Ridley, *Evolution* (Boston: Blackwell Scientific, 1993), 50.

22. Cited in Herman Bavinck, *The Philosophy of Revelation*, the Stone Lectures of 1908–9, lecture 2, available at http://www.kuyper.org/Bavinck/Stone/preface.html.

23. See, for example, Alan Fersht, *Structure and Mechanism in Protein Science* (New York: W. H. Freeman, 1999), 26–30; and Carl Branden and John Tooze, *Introduction to Protein Structure*, 2d ed. (New York: Garland, 1999), 208–19.

24. Berra, *Evolution and the Myth of Creationism*, 66.

Chapter 5

1. Emile Zuckerkandl and Linus Pauling, "Molecules as Documents of Evolutionary History," *Journal of Theoretical Biology* 8 (1965): 357–66.

2. National Academy of Sciences, *Science and Creationism: A View from the National Academy of Sciences*, 2d ed. (Washington, D.C.: National Academy Press, 1999), 19.

3. Thomas H. Jukes, "Molecular Evidence for Evolution," in *Scientists Confront Creationism*, ed. Laurie R. Godfrey (New York: W. W. Norton, 1983), 119 (emphasis in original).

4. See, for example, Peter J. Andrews, "Aspects of Hominoid Phylogeny," in *Molecules and Morphology in Evolution*, ed. Colin Patterson (Cambridge: Cambridge University Press, 1987), 28.

5. Thomas H. Jukes and Richard Holmquist, "Evolutionary Clock: Nonconstancy of Rate in Different Species," *Science* 177 (1972): 530–32.

6. Richard P. Ambler and Margaret Wynn, "The Amino Acid Sequences of Cytochromes c-551 from Three Species of *Pseudomonas*," *Biochemical Journal* 131 (1973): 485–98; and Richard P. Ambler et al., "Cytochrome c Sequence Variation among the Recognised Species of Purple Nonsulphur Photosynthetic Bacteria," *Nature* 278 (1979): 659–60.

7. Christian Schwabe and Gregory W. Warr, "A Polyphyletic View of Evolution: The Genetic Potential Hypothesis," *Perspectives in Biology and Medicine* 27 (1984): 465–78; and Christian Schwabe, "On the Validity of Molecular Evolution," *Trends in Biochemical Sciences* 11 (1986): 280–82.

8. Schwabe and Warr, "A Polyphyletic View of Evolution," 471.

9. Gabriel A. Dover, "DNA Turnover and the Molecular Clock," *Journal of Molecular Evolution* 26 (1987): 47–58.

10. Schwabe, "On the Validity of Molecular Evolution," 280.

11. For example, Herman Bavink wrote that "it is a pity that a conception which is to explain everything should itself so much need explaining." See Herman Bavinck, *The Philosophy of Revelation*, the Stone Lectures of 1908–9, lecture 2, available from http://www.kuyper.org/Bavinck/Stone/preface.html.

12. Peter E. M. Gibbs, Werner F. Witke, and Achilles Dugaiczyk, "The Molecular Clock Runs at Different Rates among Closely Related Members of a Gene Family," *Journal of Molecular Evolution* 46 (1998): 552–61.

13. Francisco J. Ayala, "Molecular Clock Mirages," *BioEssays* 21 (1999): 73

14. Michael S. Y. Lee, "Molecular Clock Calibrations and Metazoan Divergence Dates," *Journal of Molecular Evolution* 49 (1999): 389.

15. De Beer, *Atlas of Evolution*, 44.

16. Niles Eldredge, *The Triumph of Evolution and the Failure of Creationism* (New York: W. H. Freeman, 2000), 27.

17. For example, see H. H. Lane, *Evolution and Christian Faith* (Princeton: Princeton University Press, 1923); and Berra, *Evolution and the Myth of Creationism*, 19–20.

18. For example, see Mark Ridley, *Evolution* (Boston: Blackwell Scientific, 1993), 50–52; David Penny, Les R. Foulds, and Michael D. Hendy, "Testing the Theory of Evolution by Comparing Phylogenetic Trees Constructed from Five Different Protein Sequences," *Nature* 297 (1982): 297.

19. Emma C. Teeling et al.,"Microbat Paraphyly and the Convergent Evolution of a Key Innovation in Old World Rhinolophoid Microbats," *PNAS* 99 (2002): 1431–36.

20. Michael Balter, "Morphologists Learn to Live with Molecular Upstarts," *Science* 276 (1997): 1034.

21. Jason Raymond, et. al., "Whole-Genome Analysis of Photosynthetic Prokaryotes," *Science* 298 (2002): 1616–19.

22. Laura E. Maley and Charles R. Marshall, "The Coming of Age of Molecular Systematics," *Science* 279 (1998): 505–6.

23. Elizabeth Pennisi, "Charting a Genome's Hills and Valleys," *Science* 296 (2002): 1601.

24. D. H. Irwin, "Macroevolution is more than repeated rounds of microevolution," *Evol. Dev.* 2000 2 (2): 61–2.

25. National Academy of Sciences, *Science and Creationism: A View from the National Academy of Sciences*, 2d ed. (Washington, D.C.: National Academy Press, 1999), 6.

26. Ernst Mayr, *What Evolution Is* (New York: Basic Books, 2001), 272.

27. Carl Zimmer, *Evolution* (New York: HarperCollins, 2001), 104.

28. Ibid., 104.

29. Ibid.

30. Ibid.

31. Ibid.

32. Ibid., 104–5.

Chapter 6

1. Ernst Mayr, *The Growth of Biological Thought* (Cambridge: Harvard University Press/Belknap Press, 1982), 403.

2. Jonathan Weiner, "Kansas Anti-Evolution Vote Denies Students a Full Spiritual Journey," *Philadelphia Inquirer*, 15 August 1999.

3. Quoted in Steve Jones, *Darwin's Ghost* (New York: Random House, 2000), 128.

4. Charles Darwin, *The Orgin of the Species*, 6th ed. (1872; reprint. London: Collier Macmillan, 1962), 437.

5. "I should without hesitation adopt [evolution], even if it were unsupported by other facts or arguments." Darwin, *Origin*, 457.

6. Ibid.,467–72.

7. Ibid., 434–35.

8. Ibid., 437.

9. Joseph Le Conte, *Evolution: Its Nature, Its Evidences, and Its Relation to Religious Thought*, 2d ed. (New York: D. Appleton, 1891), 54.

10. H. H. Lane, *Evolution and Christian Faith* (Princeton: Princeton University Press, 1923), 32.

11. Arthur W. Lindsey, *Principles of Organic Evolution* (St. Louis: Mosby, 1952), 116.

12. Stephen Jay Gould, "The Panda's Thumb," in *The Panda's Thumb* (New York: W. W. Norton, 1980), 20.

13. Mark Ridley, *Evolution* (Boston: Blackwell Scientific, 1993), 50.

14. Gould, "The Panda's Thumb," 20.

15. Gavin De Beer, *Atlas of Evolution* (London: Nelson, 1964), 38.

16. Lane, *Evolution and Christian Faith*, 31.

17. Quoted in Michael J. Behe, *Darwin's Black Box: The Biochemical Challenge to Evolution* (New York: Free Press, 1996), 225–26.

18. Edward E. Max, "Plagiarized Errors and Molecular Genetics," *Creation/Evolution* in 1986 (XIX, p. 34), and http://www.talkorigins.org/faqs/molecular-genetics.html (last updated March 14, 2001).

19. Ridley, *Evolution*, 49.

20. See, for example, Jacques Ninio, *Molecular Approaches to Evolution* (Princeton: Princeton University Press, 1983), 79–81; Hyman Hartman, "Speculations on the Evolution of the Genetic Code," *Origins of Life and Evolution of the Biosphere* 25 (1995): 265; and Terres A. Ronneberg, Laura F. Landweber, and Stephen J. Freeland, "Testing a Biosynthetic Theory of the Genetic Code: Fact or Artifact?" *PNAS* 97 (2000): 13690–5.

21. Carl Zimmer, *Evolution* (New York: HarperCollins, 2001), 124 (emphasis added).

22. Discovery Institute, *Getting the Facts Straight: A Viewer's Guide to PBS's* Evolution (2001), 24 (emphasis added).

23. Stephen C. Stearns and Rolf F. Hoekstra, *Evolution: An Introduction* (Oxford: Oxford University Press, 2000), 127.

24. Thomas H. Jukes, "Molecular Evidence for Evolution," in *Scientists Confront Creationism*, ed. Laurie R. Godfrey (New York: W. W. Norton, 1983), 119 (emphasis in original).

25. De Beer, *Atlas of Evolution*, 44.

26. George S. Carter, *A Hundred Years of Evolution* (London: Sidgwick and Jackson, 1957), 15.

27. Joel Cracraft, "Systematics, Comparative Biology, and Creationism," in *Scientists Confront Creationism*, ed. Laurie R. Godfrey (New York: W. W. Norton, 1983), 172.

28. Niles Eldredge, *The Triumph of Evolution and the Failure of Creationism* (New York: W. H. Freeman, 2000), 146.

29. "The most puzzling event in the history of life on earth is the change from the Mesozoic, Age of Reptiles, to the Age of Mammals. It is as if the curtain were rung down suddenly on a stage where all the leading roles were taken by reptiles, especially dinosaurs in great numbers and bewildering variety, and rose again immediately to reveal the same setting but an entirely new cast, a cast in which the dinosaurs do not appear at all, other reptiles are supernumeraries, and all the leading parts are played by mammals of sorts barely hinted at in the preceding acts" (George G. Simpson in Editors of Time-Life, *Life before Man* [New York: Time-Life Books, 1972], 42).

30. George G. Simpson, *Horses* (Oxford: Oxford University Press, 1951), quoted in Richard Milton, *Shattering the Myths of Darwinism* (Rochester, Vt.: Park Street Press, 1992), 102.

31. Lane, *Evolution and Christian Faith* (Princeton: Princeton University Press, 1923), 38.

32. De Beer, *Atlas of Evolution*, 48.

33. Kenneth R. Miller, *Finding Darwin's God* (New York: Cliff Street Books, 1999), 102.

34. Douglas J. Futuyma, *Science on Trial* (New York: Pantheon Books, 1983), 80.

35. Tim M. Berra, *Evolution and the Myth of Creationism* (Stanford: Stanford University Press, 1990), 39 (emphasis in original).

36. Mark Ridley, *Evolution* (Boston: Blackwell Scientific, 1993), 56.

37. Stephen Jay Gould, "Darwinism Defined: The Difference between Fact and Theory," *Discover*, January 1987.

38. Miller, *Finding Darwin's God*, 41.

39. Futuyma, *Science on Trial*, 127.

40. Miller, *Finding Darwin's God*, 97.

41. Ibid., 100–3.

42. Herman Bavinck, *The Philosophy of Revelation*, the Stone Lectures of 1908–9, lecture 1, available from http://www.kuyper.org/Bavinck/Stone/preface.html.

43. Charles Darwin, *The Descent of Man and Selection in Relation to Sex*, 2d ed. (1871; reprint, London: John Murray, 1922), 92.

Chapter 7

1. Colin Brown, *Philosophy & the Christian Faith*, (Downers Grove: InterVarsity Press, 1968), 20.

2. Pss. 145:3, 139:6.

3. Isa. 55:8–9.

4. Job 38:2.

5. Job 40:2, 40:8.

6. Job 42:3.

7. Williston Walker, *A History of the Christian Church*, rev. ed. (New York: Charles Scribner's Sons, 1959), 438.

8. Ibid., 437.

9. Claude Welch, *Protestant Thought in the Nineteenth Century, Volume 1, 1799–1870* (New Haven: Yale University Press, 1972), 38.

10. Anders Jeffner, *Butler and Hume on Religion* (Stockholm: Aktiebolaget Trychmans, 1966), 136–50; and John Dillenberger, "The Apologetic Defence of Christianity," in *Science and Religious Belief: A Selection of Recent Historical Studies*, ed. C. A. Russel (Kent: Hodder and Stoughton, 1973), 182–86.

11. John Ray, *The Wisdom of God Manifested in the Works of the Creation*, 7th ed., corrected (1717; reprint, New York: Arno Press, 1977), 51.

12. Ibid., 51.

13. Dillenberger, "The Apologetic Defence of Christianity," 187.

14. Rom. 8:20–22.

15. William Paley, *The Principles of Moral and Political Philosophy*, 20th ed., vol. 1 (London: J. Faulder, 1814), 71.

16. Ibid.

17. William Paley, *Natural Theology: Or, Evidences of the Existence and Attributes of the Deity*, 12th ed., (London: J. Faulder, 1809), 456–57.

18. Ibid., 470.

19. Isa. 45:7.

20. Ps. 145:8–19.

21. Ps. 66:10–12; Matt. 7:14.

22. Jeffner, *Butler and Hume on Religion*, 151.

23. Charles C. Gillispie, *Genesis and Geology* (Cambridge: Harvard University Press, 1951), 209.

24. Theologian and geologist William Buckland (1784–1856); see Stephen Jay Gould, "Nonmoral Nature," in *Hen's Teeth and Horse's Toes* (New York: W. W. Norton, 1983), 32–43.

25. William Buckland, quoted in Gillispie, *Genesis and Geology*, 201.

26. Entomologist William Kirby (1759–1850); see Gould, "Nonmoral Nature."

27. Quoted in Stephen Jay Gould, *Ever Since Darwin: Reflections in Natural History* (New York: W. W. Norton, 1973), 141–46.

28. Paley, *Natural Theology*, 7.

29. David Hume, *Dialogues concerning Natural Religion*, quoted in Jeffner, *Butler and Hume on Religion*, 150.

30. David Hume, *Dialogues Concerning Natural Religion*, part x, http://www.librairie.hpg.ig.com.br/hume008.html.

31. Hume, *Dialogues*, 195–96.

32. Gillispie, *Genesis and Geology*, 195–96.

33. Quoted in Jack Morrel and Arnold Thackray, *Gentlemen of Science* (Oxford: Clarendon Press, 1981), 236.

34. Baden Powell, *The Connexion of Natural and Divine Truth* (London: J. W. Parker, 1838), quoted in Robert M. Young, *Darwin's Metaphor: Nature's Place in Victorian Culture* (Cambridge and New York: Cambridge University Press, 1985).

35. John 1:14.

36. A. N. Wilson, *God's Funeral* (New York: W. W. Norton, 1999), 129–30.

37. Philip J. Lee, *Against the Protestant Gnostics* (Oxford: Oxford University Press, 1987), 17.

38. Charles Darwin, *The Origin of Species*, 6th ed. (1872; reprint, London: Collier Macmillan, 1962), 181.

39. Gavin De Beer, *Atlas of Evolution* (London: Nelson, 1964), 44.

40. Mark Ridley, *Evolution* (Boston: Blackwell Scientific, 1993), 323. Ridley writes: "We can accept that an omnipotent, supernatural agent could create well-adapted living things: in that sense the explanation works. However, it has two defects. One is that supernatural explanations for natural phenomena are scientifically useless. The second is that the supernatural Creator is not explanatory. The problem is to explain the existence of adaptation in the world; but the supernatural Creator already possesses this property. Omnipotent beings are themselves well-designed, adaptively complex, entities. The thing we want to explain has been built into the explanation. Positing a God merely invites the question of how such a highly adaptive and well-designed thing could in its turn have come into existence."

41. Kenneth R. Miller, *Finding Darwin's God* (New York: Cliff Street Books, 1999), 97.

42. Douglas Futuyma, *Science on Trial* (New York: Pantheon Books, 1983), 46.

43. Ibid., 48.

44. Ibid., 62.

45. Stephen Jay Gould, "The Panda's Thumb," in *The Panda's Thumb* (New York: W. W. Norton, 1980), 20–21.

Chapter 8

1. Arthur W. Lindsey, *Principles of Organic Evolution* (St. Louis: Mosby, 1952), 116.

2. Douglas J. Futuyma, *Science on Trial* (New York: Pantheon Books, 1983), 199.

3. Job 39:13–18.

4. Job 38:31–39:26.

5. Job 39:19, 27–30.

6. For example: Gen. 1:1, Acts 14:15, 17:24–25, Rev. 4:11.

7. Isa. 55:9.

8. John 5:17.

9. Ps. 50:11.

10. Charles Darwin, *The Origin of Species*, 6th ed. (1872; reprint, London: Collier Macmillan, 1962), 437.

11. Gen. 3:17.

12. Rom. 5:12.

13. Gen. 1:31.

14. Ps. 19:1.

15. Isa. 6:3.

16. Jer. 10:12.

17. Job 38:4, 12, 17.

18. Ps. 104:1.

19. Ps. 104:2–3.

20. Ps. 104:5.

21. Ps. 104:10.

22. Ps. 104:14.

23. Ps. 104:15.

24. Ps. 104:19.

25. Ps. 104:24–32.

26. Col. 2:2–8.

27. Dan. 12:4,10.

28. 2 Tim. 3:7.

29. 2 Tim. 4:3–4.

30. Isa. 45:9–12.

31. Quoted in John C. Greene, *Science, Ideology, and World View* (Berkeley: University of California Press, 1981), 52.

32. Neal C. Gillespie, *Charles Darwin and the Problem of Creation* (Chicago: The University of Chicago Press, 1979), 32–33.

33. Darwin, *Origin*, 435.

34. Joseph Le Conte, *Evolution: Its Nature, Its Evidences, and Its Relation to Religious Thought,* 2d ed. rev. (New York: D. Appleton, 1891), 65–66.

35. John Rennie, "15 Answers to Creationist Nonsense," *Scientific American,* July 2002.

36. Niles Eldredge, *The Monkey Business* (New York: Washington Square, 1982), 39.

37. Paul A. Moody, *Introduction to Evolution* (New York: Harper and Row, 1970), 26.

38. Tim M. Berra, *Evolution and the Myth of Creationism* (Stanford: Stanford University Press, 1990), 66.

39. Ibid., 142.

40. Rom. 1:21.

41. Rom. 1:18–25.

42. Col. 2:8.

43. 1 Cor. 1:20.

44. Col. 3:5–8.

45. 2 Tim. 3:2–4.

46. Matt. 5:21, 27–28.

47. Matt. 5:29.

48. Matt. 5:48.

49. Rom. 14:12.

50. Gen. 6:5.

51. Ps. 40:12.

52. Jer. 17:10.

53. James 1:17 (NKJV).

54. Heb. 12:29.

55. Ps. 50:21–22.

56. Ps. 7:11–12.

57. John 6:28–29.

58. John 3:16–17.

59. Ps. 2:11.

60. Rom. 8:28.

61. Gen. 50:20.

62. Ps. 37:5–6.

Chapter 9

1. Matt. 27:35–42.
2. Luke 16:19–31.
3. Isa. 53:3–4.
4. Isa. 53:5–8.
5. Ps. 22:7–18.
6. 2 Tim. 3:16.
7. Gen. 1:31.
8. Ps. 19:1.
9. Rom. 8:20–22.
10. Job 39: 7, 14–17.
11. Anthony D. Baker, "Theology and the Crisis in Darwinism," *Modern Theology* 18, no. 2 (2002): 192.
12. Charles Darwin, *The Origin of Species*, 6th ed. (1872; reprint, London: Collier Macmillan, 1962), 77.
13. Kenneth R. Miller, *Finding Darwin's God* (New York: Cliff Street Books, 1999), 102.
14. Rom. 1:18–20.
15. Psalms 1.
16. Matt. 6:28.
17. Matt. 13:33.
18. Matthew 23:33.
19. Prov. 26:11.
20. Herman Bavinck, *The Philosophy of Revelation*, the Stone Lectures of 1908–9, lecture 4, available at http://www.kuyper.org/Bavinck/Stone/preface.html.
21. James 1:2–4.
22. Rom. 8:21.
23. Lisa J. Shawver, "Trilobite Eyes: An Impressive Feat of Early Evolution," *Science News* 105 (1974): 72.
24. Riccardo Levi-Setti, *Trilobites*, 2d ed. (Chicago: The University of Chicago Press, 1993), 29.

Chapter 10

1. Edward M. Kennedy, "Evolution Is Designed for Science Classes" (letter to the editor), *Washington Times*, 21 March 2002.
2. Iain Murray, "Scientific Boehner: The New Creationism and the Congressmen Who Support It," *The American Prospect Online*, 5 June 2002, http://www.prospect.org/webfeatures/2002/06/murray-i-06-05.html.
3. John Rennie, "15 Answers to Creationist Nonsense," *Scientific American*, July 2002.
4. Adrian Melott, "Intelligent Design Is Creationism in a Cheap Tuxedo," *Physics Today*, June 2002.
5. Andrew Oldenquist, "Biology Class Isn't Designed for Philosophy," *Columbus Dispatch*, 21 March 2002.
6. Editorial and comment, "Devolution: Creationism Proponents Take to the Trenches," *Columbus Dispatch*, 14 June 2002, 18A.
7. Editorial, "School Board Must Side with Real Science," *Dayton Daily News*, 17 June 2002.
8. Prov. 1:7.
9. Col. 2:3.

Chapter 11

1. Ray Bradbury, "A Sound of Thunder," *R is for Rocket* (Garden City, N.Y.: Doubleday, 1952.

2. Kenneth R. Miller, *Finding Darwin's God* (New York: Cliff Street Books, 1999), 237.

3. Kenneth R. Miller, "Analysis: The Discovery Institute's efforts to smear PBS's EVOLUTION series misfire badly." 25 September 2001, http://www.ncseweb.org/resources/articles/6945_km-3.pdf

4. See, for example, collected data in D. R. Knight, et. al., "Rewiring the keyboard: evolvability of the genetic code," *Nature Reviews-Genetics* 2 (2001):49–59.

5. Mark Ridley, *Evolution* (Boston: Blackwell Scientific, 1993), 53.

Appendix

1. Neil A. Campbell, *Biology*, 2d ed. (San Francisco: Benjamin Cummings, 1990), 434, quoted in Laurence Moran, "Evolution Is a Fact and a Theory," Talk.Origins archive, http://www.talkorigins.org/faqs/evolution-fact.html.

2. Douglas J. Futuyma, quoted in Moran, "Evolution Is a Fact and a Theory."

3. R. C. Lewontin "Evolution/Creation Debate: A Time for Truth," *Bioscience* 31 (1981): 559, quoted in Moran, "Evolution Is a Fact and a Theory."

4. Niles Eldredge, *The Monkey Business* (New York: Washington Square, 1982) 31–32.

5. National Academy of Sciences, *Science and Creationism: A View from the National Academy of Sciences*, 2d ed. (Washington, D.C.: National Academy Press, 1999), available at http://books.nap.edu/html/creationism/appendix.html.

6. Stephen Jay Gould, "Darwinism Defined: The Difference between Fact and Theory," *Discover*, January 1987.

7. Quoted in Theodosius Dobzhansky, *Mankind Evolving* (New Haven: Yale University Press, 1962), 347.

8. For example, see Stephen Jay Gould, "Evolution as Fact and Theory," in *Hen's Teeth and Horse's Toes* (New York: W. W. Norton, 1994), available at http://www.freethought-web.org/ctrl/gould_fact-and-theory.html.

9. Tim M. Berra, *Evolution and the Myth of Creationism* (Stanford, Calif.: Stanford University Press, 1990), 19.

10. Theodosius Dobzhansky, "Nothing in Biology Makes Sense Except in the Light of Evolution," *American Biology Teacher* 35 (March 1973), 125, available at http://www.pbs.org/wgbh/evolution/library/10/2/text_pop/l_102_01.html.

11. Mark Ridley, *Evolution* (Boston: Blackwell Scientific, 1993), 50 (ellipsis in original).

12. David Hume, *Dialogues Concerning Natural Religion*, part iv, http://www.librairie.hpg.ig.com.br/hume008.html.

13. Ridley, *Evolution*, 57.

14. Ridley, *Evolution*, 323.

Index

Adam 125
aircraft 127–8
amino acids 21–4, 27–8, 30, 48, 50, 57,
    74, 122
Anselm 82–3, 119, 125
antibiotic 60, 68
appendix 44, 71
archetype 77, 133
ATP 29, 32–3
ATP synthase 32–3

bacteria 24–5, 42, 51, 60–1
bat 39, 41, 47, 56, 69, 70, 73
Bavinck, Herman 81, 106, 114
Berra, Tim 37, 46, 48, 79, 105, 140
bird 39, 41, 45–6, 60, 67
blood immunity 55
Bradbury, Ray 130
breeding 60
Bridgewater Treatises 89
Brougham, Henry 92
Burnet, Thomas 90, 143
Butler, Joseph 37, 85–6, 91
butterfly 39

Cambrionists 86
Campbell, Ndge Plateil 136
cardiovascular system 32
Carroll, Robert 39
Carroll, Sean 75–6
Carter, George 76
catalase 27
Cheyne, George 86
chicken 57
chimpanzee 58–9
Christianity 11, 85–6, 94, 105
coccyx 44
comparative anatomy 41
    convergent evolution 46–9
    convergent phylogenies 55–8
    evidence against evolution 41–60
    hierarchical pattern 53–5, 76–7
    junk DNA 44–5
    molecular clock 50–3
    religious interpretation of 68–78
    vestigial organ 43–6
Conybeare, William 92
corn 60
Cracraft, Joel 76

creation 10
  divine 10, 67–8, 123–5
  idealization of 68–9, 71, 88, 112–3
  mechanistic 38, 90, 113
  rejection of 37, 67–81, 105, 115, 139–40
Crick, Francis 16, 130
crucifixion 89, 109–10, 112, 115
Cudworth, Ralph 118
cytochrome c 51

Darwin, Charles 9, 14, 65–71, 76–7, 81–3, 94–5, 103–4, 113, 115
  on complexity 34–5, 143, 145
  on uniformitarianism 37–8
  on the origin of life 62;
Darwin, Erasmus 85
David 101, 107, 111–2
De Beer, Gavin 45, 72, 78
De Chardin, Teilhard 139
De Maupertuis, Pierre-Louis 144
Derham, William 86–7
deism 84–5
design 45, 77, 116, 120, 123–4, 127–9
Dickens, Charles 93
divine sanction 90
DNA 15–20, 44–5, 57–8, 63, 72
  junk 44–5
DNA code 21–2, 42, 74–5, 121, 130, 132, 146
Dobzhansky, Theodosius 140
Dodson, Edward 45
Dodson, Peter 45
Donkey 99, 112
Drummond, Henry 114

e. coli 17, 19
echolocation 56
Eldredge, Niles 76–7, 104, 136
elephant 80
enzymes 15, 26–30, 48
Eve 125
evil 86–92, 97–9, 101–2, 106–8, 112
evolution 9
  convergent 41, 46–9, 78, 122
  fact of 10, 37, 136
  explanatory mechanisms for 52, 57, 59–60, 73
  large-scale 37–41, 60–1, 67
  of complexity 20–4, 28, 30, 34, 120

  small-scale 9, 24–5, 60–2, 67, 123, 138–9, 143
  variable rates of 54, 59
evolution of the gaps 13, 122

fall from grace 100–3, 106, 112
finch 65–7
fish 57, 61, 70, 78, 115
fitness 44
flower 69
fly 75
flying squirrel 47, 132
fossils 37
  conflicts with molecular clock 52–3
  evidence against evolution 38–41, 61
  religious interpretation of 78–80
  transitional 37
Franklin, Rosalind 16
frog 41, 46, 56–7, 75
Futuyma, Douglas 79, 95, 99, 136

galactose 19
Galapagos Islands 60, 65–7
Galileo 145
Gee, Henry 41
genetic code 21–2
genome 23, 25, 45, 58–9
glycolysis 28–9, 32
glucagon 33
glucose 19, 29, 33
Gnosticism 90–1, 93, 101, 119, 120, 143
God 11, 67–72, 74–7, 79–96, 98–110, 112–6, 118–9, 124, 140, 142–5, 147–150, 152
God of the gaps 149–50
Goethe, Johann 114
Gould, Stephen Jay 71, 79, 95, 138–9
GPDH 52

hemoglobin 33, 50–1, 56, 122, 128
hierarchical pattern 53–5, 76–7, 94
HIV 67
Hodge, Charles 82
homology 42, 69–70, 100
Hooker, J. D. 104
horizontal gene transfer 52, 57
horse 41, 69, 99
Hughes, Griffith 68
human 58–9, 69, 73, 78
Hume, David 91–2, 94, 106, 143, 147

Hutton, James 38
Huxley, Thomas 36–7, 64, 77

infinite regress 147–8
information 126, 128
insect 24–5, 43, 45, 51, 60
insulin 33
intelligent design 13, 45, 116–8, 120–4
Isaiah 103, 110

Jefferson, Thomas 85
Jeremiah 101
Jesus 93, 100, 107–10, 115
Job 83–4, 99, 101, 112
Joseph 108
Jukes, Thomas 76
junk DNA 44–5

Kemp, T.S. 38
Kennedy, Edward 116–7
Kepler, Johannes 66, 113
Kuhn, Thomas 117, 141

lactose 19
Lane, H. H. 70, 72, 78, 95
Laplace, Pierre 16
lateral gene transfer 52, 57
Le Conte, Joseph 10, 70, 104
Leibniz, Gottfried 98, 112–3
Lewontin, Richard 136
Lindsey, Arthur 70, 95, 98, 138
Linnaean hierarchy 53–6
Linne (Linnaeus), Carl von 53, 66–7
Lyell, Charles 38, 93
lysine 21–2

macroevolution 61
Malthus, Thomas 112
mammals 39, 41, 51, 61, 129
marsupials 46–7, 69, 123, 129, 132
Max, Edward 74
Mayr, Ernst 10, 61, 67, 95
Melott, Adrian 117
Meyer, J.B. 46
microevolution 61
Millais, John 93
Miller, Kenneth 37, 74, 79–80, 95, 113, 131–2
miracles 92
mitochondria 32, 57
molecular clock 50–3, 76

Moody, Paul 104
mosquito 56, 113
moth 67
mouse 47, 59, 60
Murray, Iain 117
mutations 24–5, 30–1, 52, 59, 72

National Academy of Sciences 51, 62, 136
natural theology 86–92, 94, 98, 112
negative control 19, 20
nerve cell 30–2
nucleotides 16, 18, 21, 27

Oldenquist, Andrew 117
OMP decarboxylase 27
orchid 71
origin of life 62–3
ostrich 99, 112
Owen, Richard 77, 133

Paine, Thomas 85
Paley, William 68–9, 87–90, 112–4
pancreas 33
paradigm of perfection 112–5, 118, 121
Paul 87, 101, 106, 112–4, 120
Pauling, Linus 16, 51
peacock 99
pentadactyl pattern 41–2, 99, 115, 130
pesticide 60, 68
phylogeny 52, 55–8, 77
pig 51
pineal gland 44
placental 46–7, 123, 129, 132
plastic nature 86–7
Plato 66
Popper, Karl 117, 151
positive control 20
potassium channel 31
Powell, Baden 92
Principle of Least Action 144
proteins 15, 23–4, 26, 30–1, 45, 57, 122
protein folding 23–4
protein synthesis 22
pseudogene 72–4
Ptolemy 11, 142
punctuated equilibrium 39–40
Pythagoreans 66

quantum mechanics 16, 123–4

Ray, John 86–7, 91
recurrent laryngeal nerve 46, 145

relaxin 51
Reimarus, Hermann 85
Rennie, John 104, 117
reptile 41, 61
rhinoceros 79
Ridley, Mark 46, 71, 74–5, 79, 95, 133, 145, 148
ribosome 21–2
RNA 17–8, 21–2
RNA hypothesis 22
RNA polymerase 17–21

Scadding, S.R. 44
Scripture 72, 83–94, 99–103, 108, 112–4, 142, 150
scurvy 73
secondary causes 90, 124
serine protease 27–8, 48
serum albumin 52
sigma factor 17–8
Simpson, George 37, 45–6, 78
sin 100–1, 107–8
snake 43, 46, 51
SOD 52
sodium channel 31
sodium-potassium pump 31
Solomon 120
species as immutable 66
spontaneous generation 138
subtilisin 48

thymus gland 44
thyroid gland 44
Tillotson, John 85
Tindal, Matthew 84
Toland, John 84
transcription 18
translation 21–2
trilobite 38, 115
tryptophan 19–20
tryptophan repressor 19–20

uniformitarianism 38

vestigial organ 43–6, 71–3, 122
virus 60, 67
vitamin C 73
Voltaire 85

Wallace, Alfred 104
Warfield, B.B. 142–3
Watson, James 16
Weiner, Jonathan 68
whale 41–2, 46–7, 71, 129
Wiedersheim, Robert 44, 122
wolf 47, 132

Zimmer, Carl 63, 75, 95
Zuckerkandl, Emile 51